Third Edition

MUSCLE TESTING

TECHNIQUES OF MANUAL EXAMINATION

LUCILLE DANIELS, M.A.

Associate Professor of Physical Therapy,
Department of Allied Medical Sciences,
School of Medicine, Stanford University

CATHERINE WORTHINGHAM, Ph.D.

Formerly, Director of Professional Education,
The National Foundation, Inc.

Format by Harold Black
Anatomical Drawings by Lorene Sigal
Test and Gait Illustrations by Harold Black and Charlene Levering

W. B. SAUNDERS COMPANY
PHILADELPHIA · LONDON · TORONTO

W. B. Saunders Company: West Washington Square
                          Philadelphia, PA 19105

                          1 St. Anne's Road
                          Eastbourne, East Sussex BN21 3UN, England

                          1 Goldthorne Avenue
                          Toronto, Ontario M8Z 5T9, Canada

Listed here is the latest translated edition of this book together with the language of the translation and the publisher.

French (3rd Edition) — La Maloine, Paris, France
German (3rd Edition) — Gustav Fischer Verlag, Stuttgart-Hohenheim, Germany
Japanese (3rd Edition) — Kyodo Isho Shuppan Sha, Tokyo, Japan
Spanish (3rd Edition) — Nueva Editorial Interamericana, S.A., de C. V., Mexico

Portuguese (3rd Edition) — Editoria Interamericana Ltda.,
                          Rio DeJaneiro, Brazil
Italian (3rd Edition) — Verduci Editore, Rome, Italy

Muscle Testing   ISBN   0-7216-2876-1

Print No.:   9   8

*The third edition of Muscle Testing is
dedicated to Marian Williams, Ph.D.,
dedicated teacher, our former co-author,
distinguished colleague, and loyal friend*

FOREWORD
to the third edition

In an era of increased recognition of the humane value of developing, restoring or maintaining function in physically impaired persons as the principal alternative to therapeutic neglect and disability, this revision of *Muscle Testing* is indeed welcome. Its accuracy, comprehensiveness and practicality have been improved, insuring a continuation of the contribution this book is making to physicians and allied health professionals, physical educators and others

Since the publication of the first edition, there has been significant increase in the need to incorporate clinical muscle testing, including its kinesiological concepts and methodologies, into the procedures required for proper patient management and for research. Methods for objective evaluation of the effects of neuromuscular impairment and measurement of changes in neuromuscular functioning must be developed in parallel with advances in therapy. Whether therapeutic efforts include drugs, surgical procedures or physical methods, orthotics or prosthetics, evaluating their effectiveness will require accurate and reliable muscle testing. Careful description of the natural history of neuromuscular disorders can probably be fully rendered only by periodic testing. Appropriately selecting and clustering patients for controlled trials of therapeutic methods depend upon application of the procedures so accurately and clearly depicted in this book.

The advanced perceptions of these authors and their unflagging determination to refine and improve the measurement of motor performance should be recognized as fundamental contributions to restorative and rehabilitative medicine. Muscle testing has already enhanced clinical decision making and led to better patient care through detection of change or lack of change in the patient's motor performance. The worldwide emphasis upon inclusion of physical restorative procedures as a regular component of medicine and the attention paid to defining and measuring the health status of large population groups will doubtless increase dependence upon this kind of test procedure or variations of it. This is a likely outcome as professionals strive to optimize and individualize patient care for many handicapped persons.

One can also expect to see growing use of muscle testing in analyzing and assessing the patient's physiological condition and in evaluating the effectiveness of care in a variety of medical conditions. The importance of clinical testing of muscle function and the relevance of interpreting and utilizing this knowledge for patient care cannot be ignored. It is therefore reassuring to see the authors' enthusiasm for an expanded use of muscle testing.

WILLIAM A. SPENCER, M.D.

Chairman, Department of Rehabilitation,
Baylor College of Medicine;
Director, Texas Institute for Rehabilitation
and Research, Houston, Texas

PREFACE
TO THE THIRD EDITION

Manual muscle tests are used to determine the extent and degree of muscular weakness resulting from disease, injury or disuse. The records obtained from the tests provide a basis for planning therapeutic procedures and periodic retesting, which can be utilized in evaluating these procedures. Muscle testing is therefore an important tool for all members of the health team dealing with the physical residual of disability.

The change in emphasis from "sick care" to "health care" and the development of comprehensive, continuous and coordinated care are resulting in increased use of muscle testing procedures. Physicians, physical therapists, occupational therapists, nurses and their supporting personnel have need for varying levels of knowledge and ability in muscle testing in their concern for preventing disability, habilitating those who have never known normal function, restoring those with disability to optimal function and maintaining the function obtained. The physical educator, although less concerned with the treatment of muscle impairment, is definitely concerned with the optimal development of the body and the prevention of disability. He must, therefore, be familiar with the details of muscle function. Testing for fair, good and normal function is an excellent device for teaching kinesiology to this group.

In this, the third edition, additional explanatory material has been added in the introduction and throughout the following sections. A number of test positions have been revised and new illustrations provided. A section has been added on gait analysis with particular emphasis on its use as a screening device and its value in obtaining a better understanding of the body as a functional unit.

There will probably always be some lack of agreement on certain principles of manual muscle testing as well as on some specific test procedures. Space has been provided throughout for notes and for the addition of any special tests which the examiner may prefer and which have particular application to the field in which he is interested.

We wish to thank the members of the faculty of the Division of Physical Therapy at Stanford University for their assistance with the new edition and in particular Barbara Kent for her detailed suggestions for revision. Our appreciation is also extended to those students in the Division who reviewed the revised material.

We are indebted, also, to many faculty members in other schools of physical therapy who responded so generously to our inquiry for suggestions.

Lucille Daniels
Catherine Worthingham

CONTENTS

II SCREENING THE AMBULATORY PATIENT FOR MUSCLE TESTING BY GAIT ANALYSIS

INTRODUCTION

The technique of manual muscle examination presented in this text is based on the work of a number of investigators. No attempt has been made to present all the tests devised for any particular muscle or muscle group. Instead, the emphasis has been placed on the testing of prime movers in relation to the principal joints of each segment of the body.

A new section concerning the use of a gait analysis for ambulatory patients has been added in this edition. Emphasis is placed on the need for an analysis as a screening device for muscle testing. The relationship between deviations in the patient's walking pattern and possible muscle weakness is outlined. The results of the gait analysis may be used to validate the test grades, and the grades may verify the results of the analysis when weakness is present.

DEVELOPMENT OF MANUAL MUSCLE TESTING

Dr. Robert W. Lovett, Professor of Orthopedic Surgery at Harvard Medical School, was the originator of the gravity tests. Janet Merrill, Director of Physical Therapeutics at Children's Hospital and the Harvard Infantile Paralysis Commission in Boston, and an early co-worker of Dr. Lovett, stated that the tests were first used in his office gymnasium in 1912. The first published article describing the procedures, which included the use of outside force, was written by Wilhelmina Wright, who assisted Dr. Lovett in his private office.

From 1912 to 1916 Dr. Lovett experimented with various types of tests and methods of recording. Shaded charts, devised by Dr. E. A. Sharpe of Buffalo, New York, were used by Lovett and Wright in 1915. These consisted of pictures of the musculature of the human body with each muscle or muscle group shaded according to the amount of strength present in the area. According to Miss Merrill, Dr. Lovett used these pictures in addition to the records of the gravity test for "visual education." They were later discarded, however, because of the difficulty of portraying graphically the wide variations in functional strength of the partially paralyzed muscles.*

---

*Letter to Lucille Daniels from Janet Merrill, January 5, 1945.

Another type of test involving the use of a spring balance was devised in 1915 by E. G. Martin, Ph.D., Assistant Professor of Physiology at Harvard, in collaboration with Dr. Lovett. A description of the test with an account of events leading to its development was published in 1916. In Dr. Lovett's book on the treatment of infantile paralysis, published in 1917, he listed the gravity tests in the section on muscle training and reprinted the article concerning the spring balance tests as a separate chapter. In this book the gravity tests were revised and included resistance with "good" and "normal" gradings, which made it possible to measure a greater range of strength and thus gave a clearer picture of the patient's condition. Tests for the trunk and neck musculature were also added. The revised tests were used throughout the country, although variations were introduced.

In 1922, Charles L. Lowman, an orthopedic surgeon in Los Angeles, worked out a numeral system for grading muscle action. It was used with the gravity tests, but covered the ranges of active motion in the joints in much greater detail, particularly when extreme muscular weakness was present. A chart for grading abdominal muscles was also devised by Dr. Lowman in 1925, and used at the Orthopaedic Hospital in Los Angeles. The final form of the chart,

LOVETT METHOD OF GRADING, 1917[1]	LOWMAN METHOD OF GRADING, 1922, NUMERALS[2]	KENDALL METHOD OF GRADING, 1936, PERCENTAGES[3]
Normal	9 (Normal) 8 (Normal minus): increase against resistance, but not quite normal	*100%* (Normal): completes arc of motion against gravity and a maximum amount of resistance several times without showing signs of fatigue
Good: when the muscle was strong enough to overcome gravity and some resistance, but not quite of normal strength	7 (Good plus): beginning power against added resistance	*80%* (Good): completes arc of motion against gravity and a medium amount of resistance several times without showing signs of fatigue, but tires quickly or is unable to complete arc of motion when done against a maximum amount of resistance
Fair: when the muscle was able to overcome gravity and could perform part of the normal movement	6 (Good): well defined control over gravity or friction	*50%* (Fair): completes the whole arc of motion against gravity, but may tire after 3 to 6 movements
Poor: when slight movement could be accomplished, but gravity could not be overcome	5 (Good minus): beginning action against gravity or friction 4 (Fair plus): beginning action of joints, but not against gravity or enough to overcome friction of table	*30%* (Greater arc of motion than 20% grading) *20%* (Poor): moves through partial arc of motion with gravity eliminated
Trace: when no movement of the limb could be accomplished, but when the muscle could be felt to contract	3 (Fair): with defined action almost up to movement of joint 2 (Fair minus): definite muscle action without muscle influence on joint 1 (Action weak): definite muscle contraction	*5%* (Trace): a contraction is felt, but there is no apparent movement of part
Totally paralyzed:	0 (Inactive): no appreciable motion	*0:* No contraction felt in muscle

[1]Lovett, R. W.: The Treatment of Infantile Paralysis. 2d ed. Philadelphia, P. Blakiston's Son & Co., 1917, p. 136.

[2]Lowman, C. L., and others: Technique of Underwater Gymnastics. Los Angeles, American Publications, Inc., 1937, p. 150.

[3]Kendall, H. O., and F. P.: Care during the Recovery Period in Paralytic Poliomyelitis. U.S. Public Health Bull. No. 242, revised 1939, p. 31.

BRUNNSTROM-DENNEN METHOD OF GRADING MOVEMENTS, 1940[4]	COMMITTEE ON AFTER-EFFECTS, NATIONAL FOUNDATION FOR INFANTILE PARALYSIS, INC. MUSCLE EXAMINATION, REVISED MARCH, 1946		
N (Normal): motion normal, taking age, sex and general muscular development of individual into consideration	100%	5 N Normal	Complete range of motion against gravity with full resistance
N−(Normal minus): normal range of motion with almost normal strength			
G+ (Good plus): same as G, but a great deal of resistance is given			
G (Good): movement against gravity and moderate resistance. At least 10 movements completed without fatigue	75%	4 G Good*	Complete range of motion against gravity with some resistance
G− (Good minus): movement against gravity and slight resistance. Range of motion practically complete. Patient performs movement at least 5 times			
F+ (Fair plus): more than 50% range of motion against gravity. Patient performs the movement at least 10 times without signs of fatigue	50%	3 F Fair*	Complete range of motion against gravity
F (Fair): 50% of normal range of motion against gravity. Patient performs motion at least 5 times			
F− (Fair minus): a small range of motion performed against gravity, but after a few repetitions fatigue sets in			
P+ (Poor plus): a range of motion of 50% or more performed with gravity eliminated, but against slight resistance. Patient performs motion at least 5 times	25%	2 P Poor*	Complete range of motion with gravity eliminated
P (Poor): at least 50% of normal range of motion performed with gravity eliminated and friction reduced. Patient repeats motion at least 5 times			
P− (Poor minus): a few degrees of motion performed when gravity is eliminated and when friction and other forces have been reduced to a minimum			
T (Trace): some tenseness or a flickering of muscle fibers can be palpated when movement is attempted	10%	1 T Trace	Evidence of slight contractility. No joint motion
0 (Zero): no contraction can be discovered in any of the muscles which perform the movement	0 0 0 Zero % S or SS Spasm C or CC Contracture		No evidence of contractility Spasm or severe spasm Contracture or severe contracture

[4]Brunnstrom, S., and Dennen, M.: Round Table on Muscle Testing. Annual Conference of American Physiotherapy Association, Federation of Crippled and Disabled, Inc., New York, 1931, pp. 1–12.

*Muscle spasm or contracture may limit range of motion. A question mark should be placed after the grading of a movement that is incomplete from this cause.

which was completed in 1929, has been published in several medical journals. It was also included in a book on underwater exercise by Lowman and Roen.

A percentage system of recording was originated in 1936 by Henry O. and Florence P. Kendall, physical therapists at the Children's Hospital School in Baltimore. This system was based on a recording range in the gravity and resistance tests from 0 to 100 per cent and introduced the element of fatigue in grading. The work of these authors, which was first published in a United States Public Health Service Bulletin, has been widely used. The first text based on this material was published in 1949.

In 1940 Signe Brunnstrom and Marjorie Dennen, physical therapists from the Institute for the Crippled and Disabled in New York City, prepared a syllabus on muscle testing. A detailed system of grading movements rather than individual muscles was outlined. This was an adaptation of the Lovett gravity and resistance tests, and included the use of fatigue as an element in grading.

In 1940, Elizabeth Kenny from Australia introduced a system for recording the presence of function, spasm and incoordination in muscles affected by poliomyelitis which was called a "muscle analysis." This was one of the many significant contributions to the treatment of patients with poliomyelitis that she made in her homeland and in this country. In 1942, Alice Lou Plastridge, Director of Physical Therapy, Georgia Warm Springs Foundation, correlated the use of the "muscle analysis" with manual muscle testing. The analysis was used primarily during the acute stage of poliomyelitis and served to supplement the strength tests in the convalescent and chronic stages.

Beginning with 1951, manual muscle testing played a vital part in the evaluation of agents designed to combat paralytic poliomyelitis. The first field trial studies were carried out to determine whether gamma globulin would protect against paralysis resulting from poliomyelitis. The project was organized under Dr. William McD. Hammon, head of the Department of Epidemiology and Microbiology of the University of Pittsburgh Graduate School of Public Health, with the support of the National Foundation for Infantile Paralysis. Muscle tests were given by physical therapists in three epidemic areas selected for the trials. Grading was done by the gravity and manual resistance techniques, using the Lovett system of grading. To assess involvement, a numerical method was devised by Dr. Jessie Wright and her associates at the D. T. Watson School of Physiatrics, Leetsdale, Pennsylvania. The muscle grades were given numerical values and each muscle or group was assigned an arbitrary factor according to its bulk. The factor multiplied by the muscle grade resulted in an "index of involvement" expressed in a percentage figure.

In the more extensive National Program for the Evaluation of Gamma Globulin in the Prophylaxis of Paralytic Poliomyelitis in 1953 the same general plan was followed, but with some changes in the grouping of the muscles tested and in the factor values assigned on the basis of muscle bulk. On the basis of individual muscle grade comparisons, it was found that physician or physical therapist examiners checking the same patients gave the same grades in approximately two-thirds to three-fourths of the tests. However, total involvement scores differed by only 3 per cent.

In the 1954 field trial of poliomyelitis vaccine 67 physical therapists participated, using the abridged form of the record previously developed.

In 1961, Smith, Iddings, Spencer and Harrington reported the development of a numerical index for clinical research in muscle testing. A detailed form was used which included the addition of + and − to the standard grades. The authors noted the importance of a numerical index of total involvement, not as a substitute for the customary muscle test, but to make it more useful in the areas of education and research. A further report by Iddings, Smith and Spencer, published the same year, discussed reliability in the clinical use of muscle testing. The results indicated that these tests can be highly reliable in spite of variation in the educational preparation of physical therapists and the use of different techniques of manual muscle testing. The average difference in grading among all physical therapists in the study was approximately 4 per cent, which compared favorably with the 3 per cent found in the more abridged form of testing in the poliomyelitis field trials.

In recent years numerous mechanical and electronic devices of varying complexity and applicability to clinical muscle testing have been developed. These provide valuable information concerning muscle function and are important for further research. However, manual muscle testing still offers a readily available and inexpensive tool for both clinical and research purposes.

A comparison of the methods of recording manual muscle tests is given in the table on pages 2 and 3.

BASIC CONSIDERATIONS IN TESTING

RELATION OF STRENGTH TO AGE AND SEX

There appears to be considerable agreement in the literature, at least in gross terms, regarding the relation of strength to age. Strength apparently increases for the first 20 years of life, remains at this level for 5 or 10 years, and then gradually decreases throughout the rest of life. Ufland, in 1933, stated that the typical curve of the changes in the strength of muscles based on age may show some deviations under the influence of occupation and constitutional types of examinees. He also stressed that changes in the muscles in aging are different for different groups of muscles and noted that the progressive decrease in strength was most pronounced in the flexor muscles of the forearm and the muscles that raise the body.

In regard to the relation of strength and sex, Galton, in 1883, stated that the strength of males increases rapidly from 2 to 19 years at a rate similar to weight, and more slowly and regularly up to 30 years, after which it declines at an increasing rate to the age of 60 years. The strength of females was found to increase at a more uniform rate from 9 to 19 years, and more slowly to 30 years, after which it falls off in a manner similar to males. In 1935, Schochrin found that women were 28 to 30 per cent weaker than men; at 40 to 45 years of age the decrease was not so great with women as men.

VALIDITY AND RELIABILITY IN TESTING

Careful observation, palpation and correct positioning are essential for validity in testing. The patient should be asked to attempt to move the part through the range of motion, if he is able to do so. The examiner should observe and note dissimilarities in the size and contour of the muscle or group of muscles being tested and the counterpart on the opposite side of the body. The contractile tissue and tendon (or tendons) should be palpated, as a lack of tension helps to identify substitution by muscles other than the prime movers. Substitution usually can be eliminated by careful positioning; if this is impossible, a notation should be made on the record. A classic example of complete substitution can occur in patients with muscular dystrophy where the prime movers may be nonfunctioning and secondary muscles perform the movement.

In considering the interpretation of a test grade it is probably unnecessary to point out the existence of variation in length and bulk of the body parts, variations in shape of like parts in different persons, age and sex differences in strength, and the ever-present psychological considerations of cooperation and willingness to put forth maximal effort which operate particularly in very small children. In view of these and other factors such as fatigue, it would be an error to assume that invariably muscles or muscle groups with the same grade have suffered an equal degree of involvement.

A factor which has been widely disregarded, not only in muscle testing but also in therapeutic exercise in general, is the considerable variation in the force which a muscle can normally exert at various points through the range of motion of the moving segment. Consideration of such "strength curves" or "isometric joint torque curves" shows that the test point in manual resistance tests is often near the weakest portion of the range. As long as the test is always done in the same manner, this will not affect its reliability, but it may have implications for functional interpretation of the grade.

THE GRADING SYSTEM

The basic grades used throughout the following pages are based on three factors:

1. The amount of resistance that can be given manually to a contracted muscle or muscle group (normal or good grade);

2. The ability of the muscle or muscle group to move a part through a complete range of motion (against gravity—a fair grade, with gravity eliminated—a poor grade); and

3. Evidence of the presence or absence of a contraction in a muscle or muscle group (slight contraction without joint motion—a trace grade, no contraction—a zero).

In addition to the use of the basic grades, a common practice is to add a plus or minus to denote:

1. A greater or lesser amount of resistance than the normal or good grades signify (slightly less resistance than the amount that can be given to a normal muscle—N—, or slight resistance at the end of the range of motion against gravity—F+);

2. A variation in the range of motion that may be graded fair or poor (range of motion can be completed with gravity eliminated and also a partial range against gravity—P+). The use of *plus* or *minus* in the resistance tests is based on a subjective decision made by the examiner. In the gravity tests, a division in the range of motion may be used which increases the objectivity of the evaluation. If less than one-half of the range is completed, the lesser grade with a plus is recorded; if more than one-half but not the full range, the higher grade with a minus is used (e.g., P+ and F− respectively for motion against gravity).

The complication of limited passive motion is of importance. The range-grade system which requires the recording of the degrees of motion with the grade is often used. For example, if the passive range in elbow flexion is limited to 90 degrees and the patient can complete this range against gravity, it is recorded as 90°/F.

NORMAL AND GOOD GRADES

The amount of resistance required for a grade of normal or good varies with the individual patient and the muscle or muscle group examined. If the muscles in the opposite extremity or side of the body from the ones being tested are known to be uninvolved, valid information can be obtained by giving resistance to each counterpart before testing the involved muscles. Otherwise, the examiner must depend upon his previous experience to make a judgment.

Resistance given at the end of the range of motion (break test) to determine good and normal grades is simpler and can be applied more quickly than resistance throughout the range of action. The patient can usually follow the request to "hold" with ease; however, be sure that he has time to establish a maximum contraction before resistance is given. Pressure should always be exerted in a direction as nearly as possible opposite to the line of pull of the muscle or muscle group being tested, and at the distal end of the segment on which it inserts.

Pain should not be elicited in a break test. Gradual, increasing pressure should be given and the patient observed closely for any evidence of discomfort or pain, and resistance discontinued if either occurs.

FAIR GRADE

The ability to raise a segment through its range of motion against gravity appears to be a fairly specific accomplishment, lying somewhere between the extremes of not being able to contract the muscle and holding the segment at the end of its range of motion against "normal" maximal resistance. Manual muscle testing in its simplest form centers around this concept, with reliance on the judgment and skill of the examiner to determine whether the muscle or group being tested is at a "fair" point in performance, or is above or below it, and to what degree.

It may appear that a direct comparison of fair grades is reasonable, since larger parts have greater muscle forces available to move them. This is true to some extent, although surprising variations exist in the ratio of the weight of the part and the maximum force normally available to lift it. For example, direct force measurements have shown that with the subject supine and the head relaxed in a sling, gravity exerts a downward force on the head which has been recorded as 9 pounds. Upward force resulting from maximum contraction of normal neck flexors may be 19 pounds, as measured by means of the sling, or a total of 28 pounds, including the force required to support the head. Thus, in this instance, the ratio of fair to normal is about 9:28, or 32 per cent. In contrast, with the subject seated the resistance of the relaxed forearm supported in a horizontal position by a test strap at the wrist may be 5 pounds and maximum elbow flexor contraction may result in an upward force of 75 pounds measured at the wrist. The ratio of these two values is 5:80, or 6.3 per cent. (Similar measurements have shown the ratio for the quadriceps to be 8:80, or 10 per cent, in some instances, and for the hip abductors 12:50, or 24 per cent.) Actually, such measurements involve force moments rather than true muscle forces and segmental weights, but if the lever lengths of downward and upward forces are kept equal, they may be ignored in estimating the ratios of the two.

The few figures cited should not be interpreted as typical of large numbers of persons, since there are undoubtedly wide variations according to age, body configuration, and other factors, as well as variations in the manner of giving the dynamometer tests. These figures are offered here to show the hazards of assigning arbitrary numerical values to the original Lovett grades; which may lead to misinterpretation. Numerals are acceptable for the recording of muscle performance only if the concepts of the grades and the tests are kept in perspective.

Direct force measurements show that the level of fair is usually relatively low, so that a vastly greater range exists between this point and normal than between this point and a trace.

A fair grade might be said to represent a *definite functional threshold* for each individual movement tested, indicating that the muscle or muscles can achieve the minimum task of moving the part upward against gravity through its range of motion. Though this ability is significant for the upper extremity, it falls far short of the functional requirements for many of the lower extremity muscles used in walking, particularly for such groups as the hip abductors, knee extensors, and plantar flexors and dorsiflexors of the foot.

POOR GRADE

The poor grade denotes the patient's ability to move a part through the range of motion with gravity eliminated. Exceptions are tests for the fingers and toes in which the weight of the parts is not significant and tests for which the gravity eliminated positions are not practical, e.g., flexion and extension of the neck. For these, a partial range may be graded poor and a full range fair.

Although considered below the functional range, muscles graded poor provide a measure of stability to a joint that is of value to the patient. It should also be noted that the identification of this level of function is important in the early stages of a disability as a muscle with a poor grade has a greater potential for an increase in strength than one that receives a lower grade of trace or zero.

TRACE AND ZERO GRADES

A trace or an absence of a muscle contraction is determined by careful observation and palpation of the tendons and the muscle bulk. An increase in tension or a flicker of movement may be more easily seen or palpated in a tendon if near the surface of the body. These should be checked first, followed by inspection and palpation of the contractile tissue. It is difficult, and sometimes impossible, to identify a minimal contraction in one of the deep muscles of the body. Usually it cannot be done unless the overlying muscles are non-functioning and the contraction of the muscle being tested is sufficient for the line of pull to be identified. Recording a trace or zero with a question mark may be indicated.

Stabilization

Manual stabilization is used in testing for adequate fixation to isolate the desired action to a specific joint. A muscle in contracting pulls on its origin as well as its insertion and with equal force. To obtain maximum muscle action, the stationary segment, which in testing is usually the site of the origin, must be fixed by muscle tension, gravitational pull or external pressure from manual stabilization. Therefore, care must be taken that muscles are not placed at a disadvantage by failure of the fixator force and consequently penalized in grading. For example, in grading glenohumeral motion, fixation of the scapula is essential.

Synergistic action refers to a contraction of all the muscles acting around a joint. These include the prime movers, the muscles which act in concert with the prime movers to define the spatial limits, and the antagonists which check or limit the movement. For example, the long finger flexors in flexing the phalanges of the fingers with maximum tension would also flex the wrist if the wrist extensors did not prevent this. In testing, the necessity for this type of muscle synergy is ordinarily eliminated by the stabilization applied by the examiner during the test.

Limitations of Manual Muscle Testing

The muscle testing methods presented in this text were designed for use in assessing the extent and degree of weakness following disorders primarily involving the contractile muscle elements, the myoneural junction and the lower motor neuron. Disorders which affect the organization of movement in the higher levels of the central nervous system, such as cerebral palsy or hemiparesis secondary to cerebral vascular accident, alter reflex activity and bring about changed states of tone in complete muscle synergies. Although muscle weakness exists, an evaluation by voluntary movements in the selected positions outlined in this book will be misleading. Methods now are available for assessment of relative degrees of weakness (hypotonia) and hypertonia in synergistic muscle groups by altering limb position and total body posture. A review of these methods, however, is not within the scope of this publication.

SCREENING TESTS

The examiners time must be conserved and the patient's fatigue considered in a detailed muscle examination; therefore, screening tests have been found to have a useful function. In one procedure, the part is placed passively by the examiner in the position used for the normal test without regard for gravity. If the patient can hold against resistance, a judgment is made by the examiner in regard to a normal or good grading. If the patient cannot hold against resistance, use of the standard tests for determining grades below good is indicated.

Another screening procedure is to combine tests for the extremities, such as checking the flexors or abductors of both shoulders simultaneously in the sitting position, and the abductors or adductors of both hips in the supine position.

With experience, the examiner will be able to devise many "quickie" tests, particularly for patients who appear to have generalized weakness. An example is the hand grasp (handshake position) in which the strength of the finger and thumb flexors can be determined by the amount and even-

ness of the pressure exerted by each phalanx against the examiner's hand. The muscles controlling wrist motion may also be tested by resistance given in this position while the examiner stabilizes the forearm. In an additonal test for the hand the examiner approximates the palmar surface of his hand to the dorsal surface of the patient's and gives resistance to all the finger and thumb extensors simultaneously.

Careful observation of patients performing ordinary activities will often provide clues to poor function and is an important part of the evaluation procedure. With health care being offered to an increasing number of geriatric patients, screening tests have become important tools for those members of the health services concerned with the welfare of these patients. With experience, the accuracy of the screening tests will be increased as the examiner develops the ability to discern not only the gross but the less obvious deviations from the normal patterns of movement. Valid judgments of the functional level may be formed without expending excessive time or unduly tiring the patients.

Gait Analysis as a Screening Device

The analysis of gait is based on meticulous observation of the ambulatory patient during standing and walking. Deviations from the normal stance of the patient that may affect gait are noted first, followed by the abnormalities in both the general and specific elements of the walking cycle. These deviations identify the areas of weakness or other factors limiting normal function. Using the clues as a guide, the examiner may then proceed to carry out such tests as are indicated. The recorded data from a gait analysis may also be used for periodically determining the degree of improvement in the basic functional activity. Movies or film strips of the patients' walking cycle are also valuable records which can be used for this purpose.

A detailed consideration of the screening procedures used in gait analysis is presented in the last section of this publication following the standard muscle tests.

NOTES CONCERNING THE TEXT

In the accompanying anatomic information, taken largely from *Gray's Anatomy of the Human Body*, muscle origins and insertions on bones are given in some detail, while connective tissue attachments are included only when they are of particular significance. The ranges of motion are approximate and based for the most part on Scott's figures.

In testing, the examiner should stand close to the patient in order that the manual force for stabilization or resistance can be given effectively and with minimal effort. It should be noted that in many of the illustrations, the examiner has stepped back from the table or is positioned on the side opposite to the one where the test ordinarily would be given in order that the full area could be photographed without obstruction or foreshortening.

Certain of the suggested testing positions should be modified for severely disabled patients. If it is necessary to use facelying or sidelying positions instead of the sitting position, for example, it should be noted on the examination record.

The following sequence in testing is suggested to avoid too frequent turning of the patient, which may not only be fatiguing but may also result in increasing the length of time necessary for the tests:

BACKLYING POSITION

Neck
    Flexion — All tests

Trunk
    Flexion — All tests
    Rotation — All tests except Poor
    Elevation of Pelvis — All tests

Hip
    Flexion — Trace and Zero
    Abduction — Poor
              Trace and Zero
    Adduction — Poor
              Trace and Zero
    Lateral Rotation — Poor
              Trace and Zero
    Medial Rotation — Poor
              Trace and Zero

Knee
    Extension — Trace and Zero

Ankle and Foot
    Plantar Flexion — Normal and Good
    Dorsiflexion and Inversion — Trace and Zero
    Inversion — Poor
              Trace and Zero
    Eversion — Poor
              Trace and Poor

Toes (4 lateral) and Hallux — All tests

Scapula
    Abduction — Normal and Good
              Fair

Shoulder
    Flexion to 90° — Trace and Zero
    Abduction to 90° — Poor
              Trace and Zero
    Horizontal Abduction — Normal and Good
              Fair

Elbow
    Flexion — Poor
              Trace and Zero
    Extension — All tests

FACELYING POSITION

Neck
Extension — All tests

Trunk
Extension — All tests

Hip
Extension — All tests except Poor

Knee
Flexion — All tests except Poor

Scapula
Adduction and Downward Rotation
— Normal and Good
Fair
Adduction — Normal and Good
Fair
Elevation — Poor
Trace and Zero
Depression — All tests

Shoulder
Extension — All tests except Poor
Horizontal Abduction — Normal and Good
Fair
Lateral Rotation — All tests
Medial Rotation — All tests

SIDELYING POSITION

Hip
Flexion — Poor
Extension — Poor
Abduction — Normal and Good
Fair
Adduction — Normal and Good
Fair

Knee
Flexion — Poor
Extension — Poor

Ankle
Plantar flexion — Poor
Trace and Zero

Foot
Inversion — Normal and Good
Fair
Eversion — Normal and Good
Fair

Shoulder
Flexion to 90° — Poor
Extension — Poor

SITTING POSITION

Trunk
Rotation—Poor

Hip
Flexion—Normal and Good
Lateral Rotation—Normal and Good
Fair
Medial Rotation—Normal and Good
Fair

Knee
Extension—Normal and Good
Fair

Ankle and Foot
Dorsiflexion and Inversion—Normal and Good
Fair
Inversion—Fair

Scapula
Abduction—Poor
Trace and Zero
Adduction and Downward Rotation—Poor
Trace and Zero
Adduction—Poor
Trace and Zero
Elevation—Normal and Good
Fair

Shoulder
Flexion to 90°—Normal and Good
Fair
Abduction to 90°—Normal and Good
Fair
Horizontal Abduction—Poor
Trace and Zero
Horizontal Adduction—Poor
Trace and Zero

Elbow
Flexion—Normal and Good
Fair

Wrist—All tests

Fingers—All tests

Thumb—All tests

STANDING POSITION

Trunk
Elevation of Pelvis—Alternate Fair

Ankle
Plantar flexion—Normal and Good
Fair

11

MUSCLE EXAMINATION

Patient's Name _____ Chart No. _____

Date of Birth _____ _____ Name of Institution _____

LEFT   RIGHT

				Examiner's Initials				
				Date				
				NECK Flexors — Sternocleidomastoideus				
				Extensor Group				
				TRUNK Flexors — Rectus abdominis				
				R. Obl. ext. abd. / L. Obl. int. abd. } Rotators { L. Obl. ext. abd. / R. Obl. ext. abd.				
				Extensors { Thoracic group / Lumbar group				
				Pelvic elev. — Quadratus lumb.				
				HIP Flexors — Iliopsoas				
				Extensors — Gluteus maximus				
				Abductors — Gluteus medius				
				Adductor group				
				Lateral rotator group				
				Medial rotator group				
				KNEE Flexors { Biceps femoris / Inner hamstrings				
				Extensors — Quadriceps femoris				
				ANKLE Plantar flexors { Gastrocnemius / Soleus				
				FOOT Invertors { Tibialis anterior / Tibialis posterior				
				Evertors { Peroneus brevis / Peroneus longus				
				TOES M. P. flexors — Lumbricales				
				I. P. flexors (1st) — Flex. digit. br.				
				I. P. flexors (2nd) — Flex. digit. l.				
				M. P. extensors { Ext. digit. l. / Ext. digit. br.				
				HALLUX M. P. flexor — Flex. hall. br.				
				I. P. flexor — Flex. hall. l.				
				M. P. extensor — Ext. hall. br.				
				I. P. extensor — Ext. hall. l.				
				GAIT:				

GRADING SYSTEM

Completes range of motion against gravity	*Completes range of motion*	*No range of motion*
N Normal—with full resistance at end of range	F Fair — against gravity	T Trace — slight contraction
G Good — with some resistance at end of range	P Poor — with gravity eliminated	0 Zero — no contraction

MUSCLE EXAMINATION

LEFT   RIGHT

				Examiner's Initials								
				Date								
			SCAPULA	Abductor	Serratus anterior							
				Elevator	Trapezius (superior) *C 34*							
				Depressor	Trapezius (inferior) *C 34*							
				Adductors	{ Trapezius (middle) *C 34*							
					{ Rhomboideus maj. & min. *C 5 Dorsal scapular.*							
			SHOULDER	Flexor	Deltoideus (anterior) *C 56 axillary*							
				Extensors	{ Latissimus dorsi *C 678 Thoracodorsal*							
					{ Teres major *Lower subscapularis C 6*							
				Abductor	Deltoideus (middle) *C 56*							
				Horiz. abd.	Deltoideus (posterior) *C 56*							
				Horiz. add.	Pectoralis major *C 5678 T1*							
				Lateral rotator group	*C 56*							
				Medial rotator group	*C 56*							
			ELBOW	Flexors	{ Biceps brachii *C 56*							
					{ Brachialis *C 56*							
				Extensor	Triceps brachii *C 78 T1*							
			FOREARM	Supinator group	*Radial C 6*							
				Pronator group	*Median C 678 T1*							
			WRIST	Flexors	{ Flex. carpi rad. *C 678 Med*							
					{ Flex. carpi uln. *C 78 T1 Ulnar*							
				Extensors	{ Ext. carpi rad. l. & br. *C 5678 Radial*							
					{ Ext. carpi uln. *C 678 radial*							
			FINGERS	M. P. flexors	Lumbricales *C 8 T1 Med + Ulnar*							
				I. P. flexors (1st)	Flex. digit. sup. *C 78 T1 Median.*							
				I. P. flexors (2nd)	Flex. digit. prof. *C 78 T1 Med + Ulnar*							
				M. P. extensor	Ext. digit. com. *C 678 Rad*							
				Adductors	Interossei palmares *C 8 T1 Ulnar*							
				Abductors	Interossei dorsales *C 8 T1 Ulnar*							
				Abductor digiti minimi	*C 8 T1 Ulnar*							
				Opponens digiti minimi	*C 8 T1 Ulnar*							
			THUMB	M. P. flexor	Flex. poll. br. *C 8 T1, C 67 Ulnar + Median*							
				I. P. flexor	Flex. poll. l. *C 8 T1 Median.*							
				M. P. extensor	Ext. poll. br. *C 678 Radial*							
				I. P. extensor	Ext. poll. l. *C 678 Radial*							
				Abductors	{ Abd. poll. br. *C 678 T1 Median*							
					{ Abd. poll. l. *C 678 Radial*							
				Adductor pollicis	*C 8 T1 Ulnar*							
				Opponens pollicis	*C 678 T1 Median*							
			FACE:									

Additional data:

I MUSCLE TESTS

NECK FLEXION

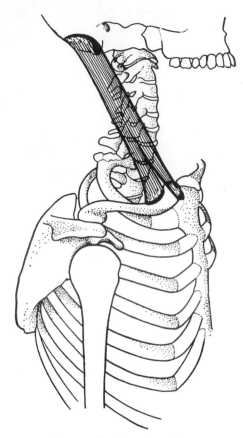

Sternocleidomastoideus

Range of Motion:

Cervical spine flexes to just beyond point where convexity is straightened. (Greatest part of motion takes place in atlanto-occipital joint.)

Factors Limiting Motion:

1. Tension of posterior longitudinal ligament, ligamenta flava, and interspinal and supraspinal ligaments
2. Tension of posterior muscles of neck
3. Apposition of lower lips of vertebral bodies anteriorly with surfaces of subjacent vertebrae
4. Compression of intervertebral fibrocartilages in front

Fixation:

1. Contraction of anterior abdominal muscles
2. Weight of thorax and upper extremities

PRIME MOVER

MUSCLE	ORIGIN	INSERTION
Sternocleidomastoideus N: Spinal accessory and (C2, 3)	*Sternal head.* a. Upper part of anterior surface of manubrium sterni *Clavicular head:* a. Superior border and anterior surface of medial third of clavicle	a. Lateral surface of mastoid process from apex to superior border b. By thin aponeurosis into lateral half of superior nuchal line of occipital bone

Accessory Muscles

Longus capitis	Scalenus medius
Longus colli	Scalenus posterior
Scalenus anterior	Rectus capitis anterior

Infrahyoid group

NECK FLEXION

NORMAL AND GOOD

Backlying.
Stabilize lower thorax.
Patient flexes cervical spine through range of motion. Resistance is given on forehead.

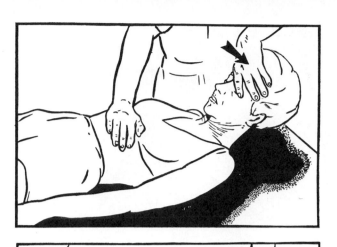

NORMAL AND GOOD

If there is a difference in strength of the two Sternocleidomastoideus muscles, they may be tested separately by rotation of head to one side and flexion of neck. Resistance is given above ear. (Test for left side illustrated.)

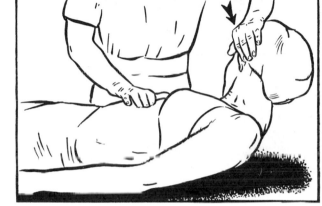

FAIR AND POOR

Backlying.
Stabilize lower thorax.
Patient flexes cervical spine through full range of motion for fair grade and through partial range for poor.

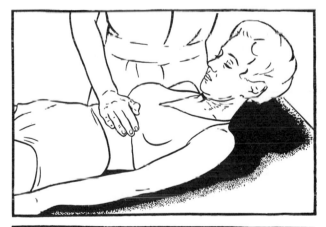

TRACE AND ZERO

The Sternocleidomastoideus muscles may be palpated on each side of neck as patient attempts to flex.

★ ★ ★ ★ ★

Note: If the accessory muscles are weak, the contraction of strong Sternocleidomastoideus muscles will increase rather than decrease the convexity of the cervical spine. The head can be raised but will be rotated posteriorly, chin up ("turtle neck position").

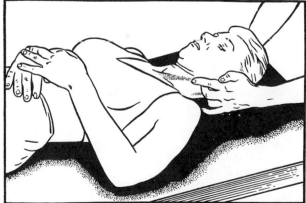

NECK EXTENSION

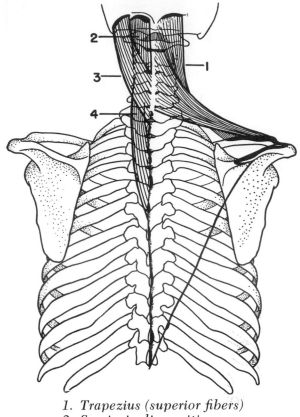

Range of Motion:

Cervical spine extends until head contacts dorsal muscle mass of upper trunk.

Factors Limiting Motion:

1. Tension of anterior longitudinal ligament of spine
2. Tension of ventral neck muscles
3. Approximation of spinous processes

Fixation:

1. Contraction of spinal extensor muscles of thorax and depressor muscles of scapulae and clavicles
2. Weight of trunk and upper extremities

1. *Trapezius (superior fibers)*
2. *Semispinalis capitis*
3. *Splenius capitis*
4. *Splenius cervicis*

PRIME MOVERS

MUSCLE	ORIGIN	INSERTION
Trapezius (superior fibers) N: Spinal accessory and (C3, 4)	a. External occipital protuberance and medial third of superior nuchal line b. Upper part of ligamentum nuchae	a. Dorsal border of lateral third of clavicle
Semispinalis capitis N: Dorsal rami of spinal nerves	a. Transverse processes of first 6 or 7 thoracic and seventh cervical vertebrae b. Articular processes of fourth, fifth and sixth cervical vertebrae	a. Between superior and inferior nuchal lines of occipital bone
Splenius capitis N: Dorsal rami of middle and lower cervical nerves	a. Caudal half of ligamentum nuchae b. Spinous processes of seventh cervical vertebra and first 3 or 4 thoracic vertebrae	a. Occipital bone just caudal to the lateral third of superior nuchal line b. Mastoid process of temporal bone
Splenius cervicis N: Dorsal rami of middle and lower cervical nerves	a. Spinous processes of third to sixth thoracic vertebrae	a. Transverse processes of upper 2 or 3 cervical vertebrae

(Continued on page 20)

NECK EXTENSION

NORMAL AND GOOD

Facelying with neck in flexion.
Stabilize upper thoracic area and scapulae.
Patient extends cervical spine through range of motion. Resistance is given on occiput.

Note: Extensor muscles on right may be tested by rotation of head to right with extension, and vice versa.

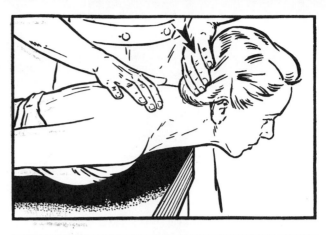

FAIR AND POOR

Facelying with neck flexed.
Stabilize upper thoracic area and scapulae.
Patient extends cervical spine through full range of motion for fair grade or through partial range for poor.

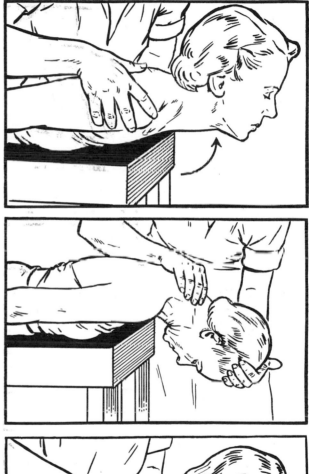

TRACE AND ZERO

Facelying.
A trace may be determined by observation and palpation of the muscles of the dorsal area of the neck. (Test may be given with head resting on table.)

Note: Be sure patient completes full range of motion of neck extension. Back muscles may contract and lift upper trunk from table, giving the appearance of extension in cervical region.

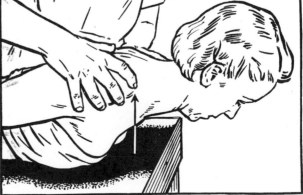

PRIME MOVERS

MUSCLE	ORIGIN	INSERTION
Erector spinae (Cervical and capitate sections not illustrated.) N: Adjacent spinal nerves Iliocostalis cervicis:	a. Angles of third to sixth ribs	a. Posterior tubercles of transverse processes of fourth to sixth cervical vertebrae
Longissimus capitis:	a. Transverse processes of first 4 or 5 thoracic vertebrae b. Articular processes of last 3 or 4 cervical vertebrae	a. Posterior margin of mastoid process
Longissimus cervicis:	a. Transverse processes of first 4 or 5 thoracic vertebrae	a. Posterior tubercles of transverse processes of second to sixth cervical vertebrae
Spinalis capitis:	(Inseparably connected with Semispinalis capitis) a. Tips of transverse processes of upper sixth or seventh thoracic and seventh cervical vertebrae b. Articular processes of last 3 cervical vertebrae	a. Between superior and inferior nuchal lines of occipital bone
Spinalis cervicis:	a. Caudal portion of ligamentum nuchae b. Spinous process of seventh cervical vertebra c. Sometimes from spinous processes of first and second thoracic vertebrae	a. Spinous process of axis b. Occasionally into spinous processes of second and third cervical vertebrae
Semispinalis cervicis N: Dorsal rami of spinal nerves	a. Transverse processes of first 5 or 6 thoracic vertebrae	a. Spinous processes of second to fifth cervical vertebrae

Accessory Muscles
Multifidus
Obliquus capitis superior and inferior
Rectus capitis posterior major and minor
Levator scapulae

FOR NOTES:

TRUNK FLEXION

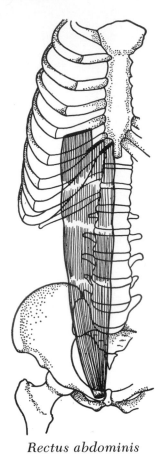

Rectus abdominis

Range of Motion:

In backlying position, flexion of thorax on pelvis is possible until scapulae are raised from table. Motion takes place primarily in thoracic spine. (Trunk is carried through remainder of range to sitting position by reverse action of hip flexors with abdominal muscles acting as fixators.)

Factors Limiting Motion:

1. Tension of posterior longitudinal ligament, ligamenta flava, and interspinal and supraspinal ligaments
2. Tension of spinal extensor muscles
3. Apposition of caudal lips of vertebral bodies anteriorly with surfaces of subjacent vertebrae
4. Compression of ventral part of intervertebral fibrocartilages
5. Contact of last ribs with abdomen

Fixation:

1. Reverse action of hip flexor muscles
2. Weight of legs and pelvis

PRIME MOVER

Muscle	Origin	Insertion
Rectus abdominis N: Intercostal nerves (7–12)	a. Crest of pubis b. Ligaments covering ventral surface of symphysis pubis	a. By 3 portions into cartilages of fifth, sixth and seventh ribs

Accessory Muscles
Obliquus internus abdominis
Obliquus externus abdominis (reverse action)

TRUNK FLEXION

NORMAL

Backlying with hands behind neck.
Stabilize legs firmly.
Patient flexes thorax on pelvis through range of motion.
(If hip flexor muscles are weak, stabilize pelvis. A curl up is emphasized and flexion is possible until scapulae are raised from table. Tests for neck flexion should precede those for trunk flexion.)

Don't hold legs down until scapulae have cleared.

TRUNK FLEXION

GOOD

Backlying with arms at sides.
Stabilize legs firmly.
Patient flexes thorax on pelvis through range of motion.
(If hip flexor muscles are weak, stabilize pelvis. Flexion is possible until scapulae are raised from table.)

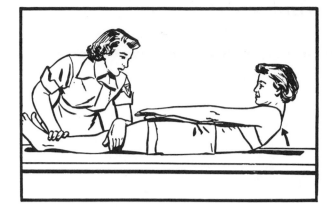

FAIR

Backlying with arms at sides.
Stabilize legs firmly.
Patient flexes thorax on pelvis through partial range of motion. Head, tips of shoulders and cranial borders of scapulae should clear table with inferior angle remaining in contact with table.
(If hip flexor muscles are weak, stabilize pelvis.)

POOR

Backlying with arms at sides
Patient flexes cervical spine. Caudal portion of thorax is depressed, and pelvis is tilted until the lumbar area of spine is flat on table.
Palpation will help to determine smoothness of contraction.

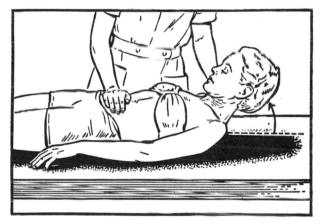

TRACE AND ZERO

Backlying.
A slight contraction may be determined by palpation over anterior abdominal wall as patient attempts to cough (also during rapid exhalation or as patient attempts to lift head).
Observe deviation of umbilicus. Cranial movement indicates stronger contraction of upper section of muscle, and caudal movement, stronger contraction of lower section (not illustrated.)

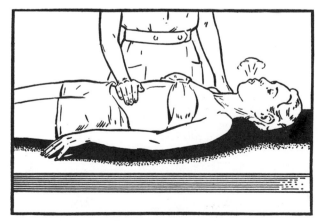

TRUNK ROTATION

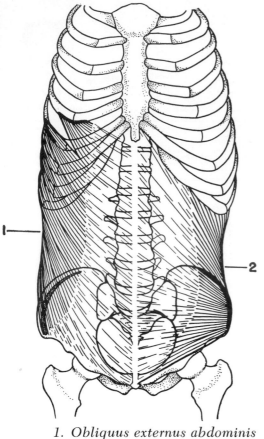

1. Obliquus externus abdominis
2. Obliquus internus abdominis

Range of Motion:

In backlying position, rotation of thorax is possible until scapula on side of forward shoulder is raised from table.

Factors Limiting Motion:

1. Tension of annulus fibrosus between vertebrae
2. Tension of oblique abdominal muscles on side opposite those being tested
3. In thoracic area, tension of costovertebral ligaments
4. In lumbar area, interlocking of articular facets. (Rotation negligible.)

Fixation:

Reverse action of hip flexor muscles.

PRIME MOVERS

MUSCLE	ORIGIN	INSERTION
Obliquus externus abdominis N: Intercostals (8–12) Iliohypogastric Ilioinguinal	a. Eight digitations from external surfaces and inferior borders of lower eight ribs	a. Anterior half of iliac crest b. Aponeurosis to pubic tubercle and pectineal line in middle, interlaces with aponeurosis of opposite muscle forming linea alba, extending from xiphoid process to symphysis pubis
Obliquus internus abdominis N: Intercostals (8–12) branch from iliohypogastric and sometimes ilio-inguinal nerves	a. Lateral half of inguinal ligament b. Anterior two thirds of iliac crest c. Posterior layer of the thoracolumbar fascia	a. Crest of pubis and medial part of pectineal line b. Linea alba c. Cartilages of seventh, eighth, and ninth ribs d. Inferior borders of cartilages of the last three or four ribs

Accessory Muscles

Latissimus dorsi Semispinalis	Multifidus Rotatores	Rectus abdominis (combined trunk rotation and flexion)

TRUNK ROTATION

NORMAL

Backlying with hands behind neck.
Stabilize legs firmly.
Patient rotates and flexes thorax to one side.
 Repeat to opposite side.
Test for left Obliquus externus abdominis
 and right Obliquus internus abdominis is
 shown in illustration. Rotation to left is
 brought about by opposite muscles.
(If hip flexor muscles are weak, stabilize
 pelvis as in "Fair" test. Upper thorax
 should be lifted from table with rotation.)

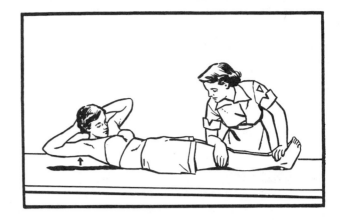

GOOD

Backlying with arms at sides.
Stabilize legs firmly.
Patient rotates and flexes thorax to one side.
 Repeat to opposite side.
(If hip flexor muscles are weak, stabilize
 pelvis as in "Fair" test.)

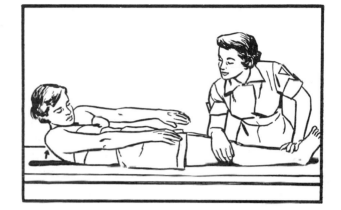

FAIR

Backlying with hands on opposite shoulders.
Stabilize pelvis.
Patient rotates thorax until scapula on side
 of forward shoulder is raised from table.
 Repeat with rotation to opposite side.

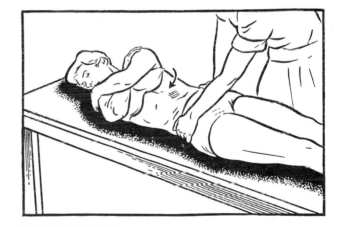

POOR

Sitting with arms relaxed at sides.
Pelvis stabilized.
Patient rotates thorax. Repeat with rotation
 to opposite side.

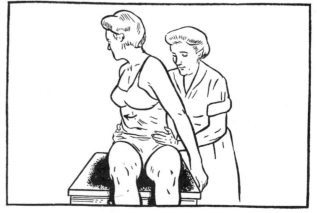

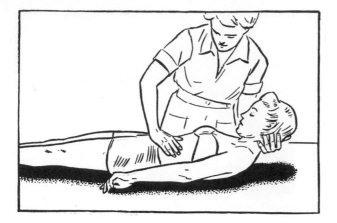

TRUNK ROTATION

TRACE AND ZERO

Examiner palpates muscles as patient attempts to approximate thorax on left and pelvis on right. Repeat on opposite side.

★ ★ ★ ★ ★

Note: Observe deviation of umbilicus, which will move toward strongest quadrant if there is a difference in strength of opposing oblique muscles (not illustrated).

FOR NOTES:

TRUNK EXTENSION

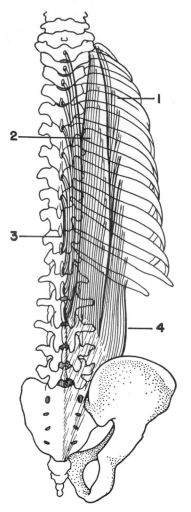

Range of Motion:

Thoracic spine extends only to approximately a straight line. Lumbar spine extends freely.

Factors Limiting Motion:

1. Tension of anterior longitudinal ligament of spine
2. Tension of anterior abdominal muscles
3. Contact of spinous processes
4. Contact of caudal articular margins with laminae

Fixation:

1. Contraction of Gluteus maximus and hamstring muscles
2. Weight of pelvis and legs

Erector spinae:

1. *Iliocostalis thoracis*
2. *Longissimus thoracis*
3. *Spinalis thoracis*
4. *Iliocostalis lumborum*

PRIME MOVERS

MUSCLE	ORIGIN	INSERTION
Erector spinae N: Adjacent spinal nerves Iliocostalis thoracis	a. Upper borders of angles of last 6 ribs	a. Superior borders of angles of first 6 ribs b. Transverse process of seventh cervical vertebra
Longissimus thoracis	a. Common tendon of sarcospinalis b. Transverse processes of lumbar vertebrae c. Anterior layer of lumbocostal fascia	a. Tips of transverse processes of all thoracic vertebrae b. Last 9 or 10 ribs between their tubercles and angles

TRUNK EXTENSION

NORMAL AND GOOD
(Extension of lumbar spine)

Facelying.

Stabilize pelvis.

Patient extends lumbar spine until caudal part of thorax is raised from table. Resistance is given on caudal portion of thoracic area.

(Tests for neck extension should precede those for trunk extension.)

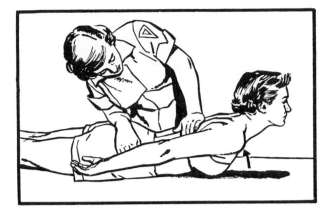

NORMAL AND GOOD
(Extension of thoracic spine)

Facelying.

Stabilize pelvis and lower part of thorax.

Patient extends thoracic spine to horizontal position. Resistance is given on cranial portion of thorax. (A pad can be placed under caudal portion of thorax if a greater range of motion is needed.)

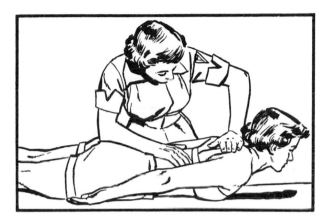

FAIR
(Extension of thoracic and lumbar spine)

Facelying.

Stabilize pelvis.

Patient extends thoracic and lumbar spine through range of motion.

POOR
(Extension of thoracic and lumbar spine)

Facelying.

Stabilize pelvis.

Patients completes partial range of motion. (Not illustrated.)

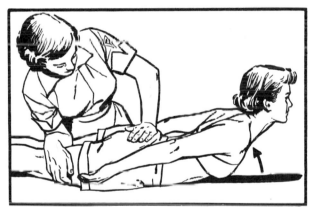

TRACE AND ZERO

Facelying.

Examiner palpates spinal extensor muscles to determine presence and degree of contraction as patient attempts to raise trunk.

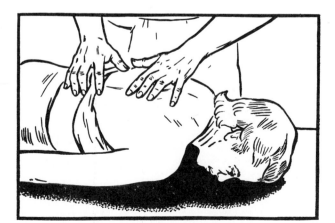

PRIME MOVERS

MUSCLE	ORIGIN	INSERTION
Erector spinae (*Continued*) Spinalis thoracis	a. Spinous processes of first 2 lumbar and last 2 thoracic vertebrae	a. Spinous processes of first 4 to 8 thoracic vertebrae
Iliocostalis lumborum	a. Common tendon: middle crest of sacrum, spinous processes of lumbar and eleventh and twelfth thoracic vertebrae, supraspinal ligament, posterior portion of inner lip of iliac crest, and lateral crest of sacrum	a. Inferior borders of angles of last 6 or 7 ribs
Quadratus lumborum (Illustrated on p. 32) N: (Th12, L1, 2)	a. Iliolumbar ligament and adjacent 5 cm. of iliac crest	a. Medial half of inferior border of last rib b. Apices of transverse processes of first 4 lumbar vertebrae

Accessory Muscles
Semispinales   Rotatores
Multifidus

FOR NOTES:

ELEVATION OF PELVIS

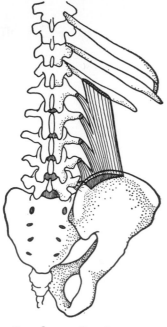

Quadratus lumborum

Range of Motion:

In standing position pelvis may be raised on one side until foot is well clear of floor. (Reverse action of Quadratus lumborum.)

Factors Limiting Motion:

1. Tension of spinal ligaments on opposite side
2. Contact of iliac crest with thorax

Fixation:

Contraction of spinal extensor muscles to fix thorax

PRIME MOVERS

MUSCLE	ORIGIN	INSERTION
Quadratus lumborum (reverse action) N: (Th12, L1, 2)	a. Iliolumbar ligament and adjacent 5 cm. of iliac crest	a. Medial half of inferior border of last rib b. Small tendons into apices of transverse processes of first 4 lumbar vertebrae
Additional portion occasionally present:	a. Transverse processes of last 3 or 4 lumbar vertebrae	a. Inferior margin of last rib
Iliocostalis lumborum (reverse action; illustrated on p. 28) N: Adjacent spinal nerves	a. Common tendon: middle crest of sacrum, spinous processes of lumbar and eleventh and twelfth thoracic vertebrae, supraspinal ligament, posterior portion of inner lip of iliac crest, and lateral crest of sacrum	a. Inferior borders of angles of last 6 or 7 ribs

Accessory Muscles

Obliquus externus abdominis (lateral fibers)
Obliquus internus abdominis (lateral fibers; reverse action)

Latissimus dorsi (with arms fixed, reverse action)
Hip abductor muscles (on opposite side; reverse action)

ELEVATION OF PELVIS

NORMAL AND GOOD

Backlying (or Facelying) with lumbar area
of spine in moderate extension. Patient
grasps edge of table to stabilize thorax. (If
arm and shoulder muscles are weak, an
assistant should stabilize thorax.)
Patient draws pelvis toward thorax on one
side. Resistance is given above ankle
joint.

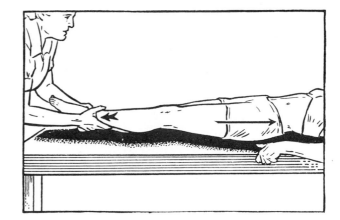

FAIR AND POOR

Backlying with legs straight and with lum-
bar area of spine in moderate extension.
Patient may grasp side of table to stabilize
thorax (not shown in picture).
Patient draws pelvis upward toward thorax.
Slight resistance is given for a fair grade.
Completion of range is graded poor.

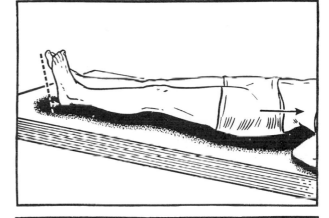

FAIR
(Alternate)

Standing position.
Stabilize thorax.
Patient lifts pelvis toward thorax through
range of motion.

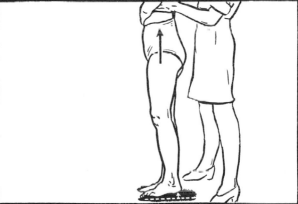

TRACE AND ZERO

As patient attempts to draw pelvis cranial-
ward, a contraction of Quadratus lumborum may be determined by deep palpation in lumbar area under lateral edge of
Erector spinae.

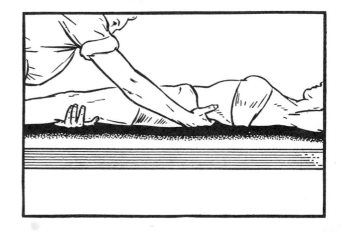

HIP FLEXION

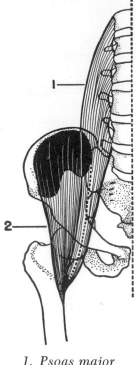

1. *Psoas major*
2. *Iliacus*

Range of Motion:

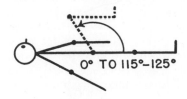

0° TO 115°–125°

Factors Limiting Motion:

1. With knee flexed, contact of thigh on abdomen
2. With knee extended, tension of hamstring muscles

Fixation:

1. Contraction of anterior abdominal muscles to fix lumbar spine and pelvis
2. Weight of trunk

PRIME MOVERS

MUSCLE	ORIGIN	INSERTION
Psoas major N: (L2, 3)	a. Transverse processes of all lumbar vertebrae b. Sides of bodies of last thoracic and all lumbar vertebrae and corresponding intervertebral fibrocartilages	a. Lesser trochanter of femur
Iliacus N: Femoral (L2, 3)	a. Superior two thirds of iliac fossa b. Inner lip of iliac crest c. Base of sacrum	a. Lateral side of tendon of psoas major b. Body of femur distal to lesser trochanter

Accessory Muscles

Rectus femoris   Pectineus
Sartorius   Adductor brevis
Tensor fasciae latae   Adductor longus
Adductor magnus (oblique fibers)

HIP FLEXION

NORMAL AND GOOD

Sitting with legs over edge of table.
Stabilize pelvis.
Patient flexes hip through last part of range
of motion. Resistance is given proximal to
knee joint.

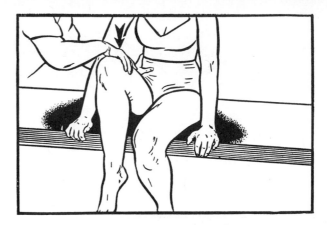

FAIR

Sitting with legs over edge of table.
Stabilize pelvis.
Patient flexes hip through last part of range
of motion.

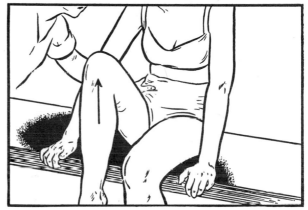

POOR

Sidelying with upper leg supported. Trunk,
pelvis and legs straight.
Stabilize pelvis.
Patient flexes hip through range of motion.
Knee is allowed to flex to prevent ham-
string tension.

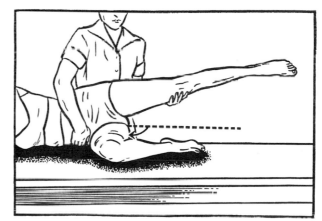

TRACE AND ZERO

Backlying with leg supported. It may be
possible to detect contraction in Psoas
major just distal to inguinal ligament on
medial side of Sartorius.

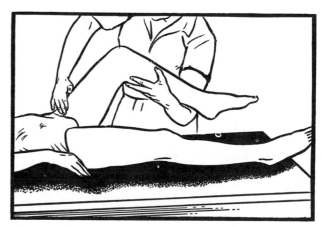

HIP FLEXION

Note: Substitution by Sartorius in hip flexion will cause lateral rotation and abduction of thigh. Muscle may easily be seen and palpated near its origin during the motion.

Note: Substitution by Tensor fasciae latae in hip flexion causes medial rotation and abduction of the thigh. Muscle may be seen and palpated at its origin. (Not illustrated.)

FOR NOTES:

HIP EXTENSION

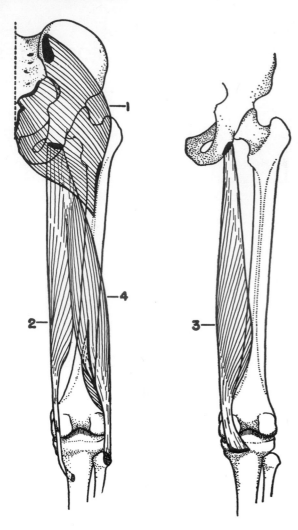

Range of Motion:

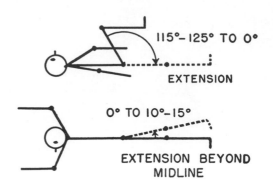

115°–125° TO 0°

EXTENSION

0° TO 10°–15°

EXTENSION BEYOND MIDLINE

Factors Limiting Motion:

    1. Tension of iliofemoral ligament
    2. Tension of hip flexor muscles

Fixation:

    1. Contraction of Iliocostalis lumborum and Quadratus lumborum muscles
    2. Weight of trunk

1. Gluteus maximus
2. Semitendinosus
3. Semimembranosus
4. Biceps femoris (long head)

PRIME MOVERS

MUSCLE	ORIGIN	INSERTION
Gluteus maximus N: Inferior gluteal (L5, S1, 2)	a. Posterior gluteal line and lateral lip of iliac crest superior and dorsal to line b. Posterior surface of lower part of sacrum and side of coccyx c. Posterior surface of sacrotuberous ligament	a. Iliotibial band of fascia lata over greater trochanter b. Gluteal tuberosity
Semitendinosus N: Sciatic (tibial) L4, 5, S1, 2)	a. Inferior and medial impression on ischial tuberosity	a. Anteromedial surface of tibia at proximal end of shaft

(Continued on page 40)

HIP EXTENSION

NORMAL AND GOOD

Facelying with legs extended.
Stabilize pelvis.
Patient extends hip through range of motion.
Resistance is given proximal to knee joint.

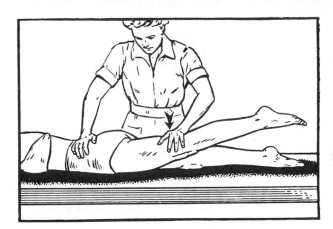

NORMAL AND GOOD
(Test for isolation of Gluteus maximus)

Facelying with knee flexed.
Stabilize pelvis.
Patient extends hip, keeping knee flexed to
    decrease action of hamstrings. Resistance
    is given proximal to knee joint.
Range of motion will be more limited than
    in position above, owing to tension in the
    Rectus femoris.

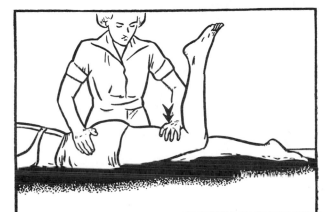

FAIR

Facelying with legs extended.
Stabilize pelvis.
Patient extends leg through range of motion.

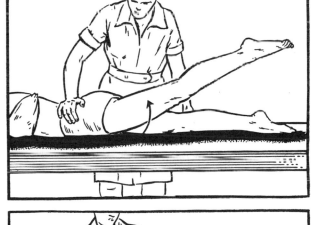

POOR

Sidelying with hip flexed, knee extended
    and upper leg supported.
Stabilize pelvis.
Patient extends hip through range of motion.
(Knee may be flexed for fair and poor to
    isolate the action of the Gluteus maxi-
    mus.)

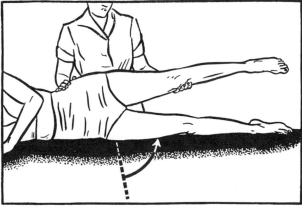

HIP EXTENSION

TRACE AND ZERO

Facelying.

Contraction of Gluteus maximus will result in narrowing of gluteal crease. Lower and upper sections of muscle should be palpated.

★ ★ ★ ★ ★

Note: Patient may lift pelvis and support leg with hamstrings, raising leg from table by extending lumbar spine. Examiner must be certain that pelvis is stable and movement takes place in hip joint.

PRIME MOVERS

Muscle	Origin	Insertion
Semimembranosus N: Sciatic (tibial) (L, 5, S1, 2)	a. Superior and lateral impression on ischial tuberosity	a. Horizontal groove on poster-omedial aspect of medial condyle of tibia b. Fibrous expansion into fascia covering Popliteus, the tibial collateral ligament and the fascia of the leg
Biceps femoris (long head) N: Sciatic (tibial) (S1, 2, 3)	*Long head:* a. Distal and medial impression on ischial tuberosity	a. Lateral side of head of fibula b. Slip to lateral condyle of tibia

Thomas Test — leg flat before small of back
I do before prone   uses off table. Clue — back arch if legs flat
test for glu max.

Eli Test — for tight of rectus femoris. Because
max is tested when knee is flexed. If knee bends to
90° w/out back rising, test is −. Knee may have to be
kept extended. If +, put pillow under abdomen.

G max test — knee passive. Don't want hams. During
passive range — extensors must be relaxed.
   poor — on side

HIP ABDUCTION

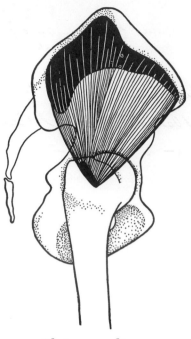

Gluteus medius

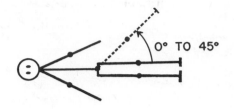

0° TO 45°

Factors Limiting Motion:

    1. Tension of distal band of iliofemoral ligament and pubocapsular ligament
    2. Tension of hip adductor muscles

Fixation:

    1. Contraction of lateral abdominal muscles and Latissimus dorsi
    2. Weight of trunk

PRIME MOVER

MUSCLE	ORIGIN	INSERTION
Gluteus medius N: Superior gluteal (L4, 5, S1)	a. Outer surface of ilium between iliac crest and posterior gluteal line dorsally and anterior gluteal line ventrally	a. Oblique ridge on lateral surface of greater trochanter

Accessory Muscles

Gluteus minimus
Tensor fasciae latae
Gluteus maximus (upper fibers)

HIP ABDUCTION

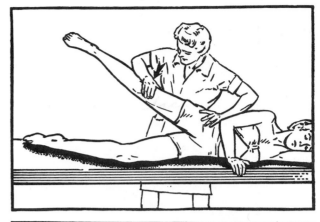

NORMAL AND GOOD

Sidelying with leg slightly extended beyond
   midline. Lower knee flexed for balance.
Stabilize pelvis.
Patient abducts leg through range of motion
   without lateral rotation of the hip.
Resistance is given proximal to knee joint.

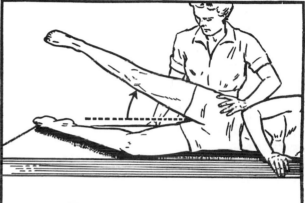

FAIR

Sidelying with leg slightly extended beyond
   midline. Lower knee flexed for balance.
Stabilize pelvis.
Patient abducts leg through range of motion.

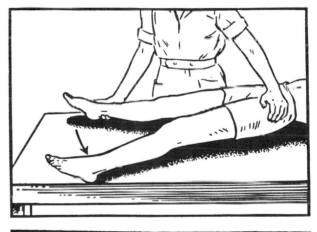

POOR

Backlying with legs extended.
Stabilize pelvis.
Patient abducts leg through range of motion
   without allowing leg to rotate.

TRACE AND ZERO

Fibers of the Gluteus medius may be found
   on lateral aspect of ilium above greater
   trochanter of femur.

HIP ABDUCTION

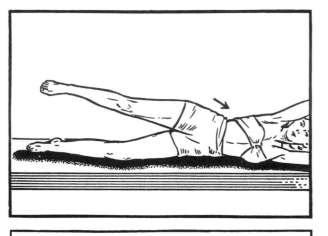

Note: Patient may bring pelvis to thorax by strong contraction of lateral trunk muscles, thereby lifting leg through partial abduction. Examiner must stabilize pelvis to make sure motion takes place in hip joint.

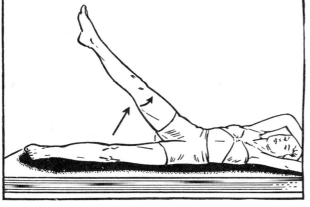

Note: Lateral rotation at the hip should be eliminated, or hip flexors may substitute for Gluteus medius. Flexion of the hip allows substitution by the Tensor fasciae latae (not illustrated).

FOR NOTES:

HIP ADDUCTION

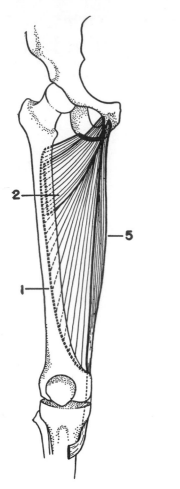

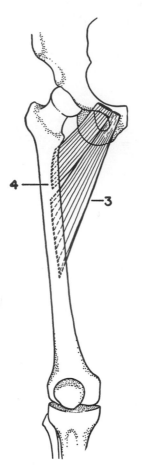

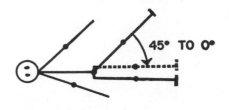

Range of Motion:

45° TO 0°

Factors Limiting Motion:

1. Contact with opposite leg
2. When hip is flexed, tension of ischio-femoral ligament

Fixation:

Weight of trunk

1. *Adductor magnus*
2. *Adductor brevis*
3. *Adductor longus*
4. *Pectineus*
5. *Gracilis*

PRIME MOVERS

MUSCLE	ORIGIN	INSERTION

Adductor magnus
N: Obturator
    (L3, 4) and branch
    from sciatic
	a. Outer margin of inferior surface of tuberosity of ischium
	b. Inferior ramus of ischium
	c. Anterior surface of inferior ramus of pubis
		a. Whole length of linea aspera and its medial prolongation
		b. Adductor tubercle

Adductor brevis N: Obturator (L3, 4)	a. Outer surface of inferior ramus of pubis	a. Distal two thirds of line leading from lesser trochanter to linea aspera, and into the proximal part of linea aspera

(Continued on page 48.)

HIP ADDUCTION

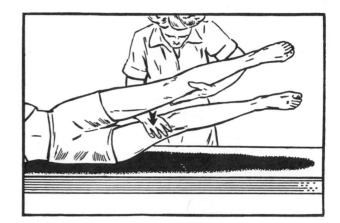

NORMAL AND GOOD

Sidelying with leg resting on table and upper leg supported in approximately 25 degrees of abduction.

Patient adducts leg until it contacts upper leg.

Resistance is given proximal to knee joint.

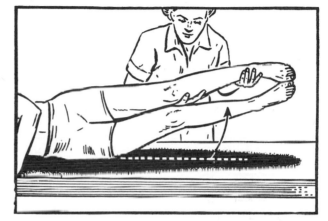

FAIR

Sidelying with leg resting on table and upper leg supported in approximately 25 degrees of abduction.

Patient adducts leg until it contacts upper leg.

POOR

Backlying with leg in 45 degrees of abduction.

Stabilize pelvis.

Patient adducts leg through range of motion without allowing rotation of hip.

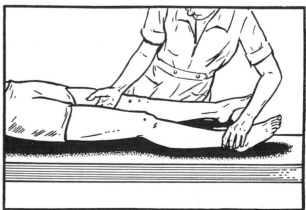

TRACE AND ZERO

Contraction of fibers of adductor muscles may be palpated on medial aspect of thigh.

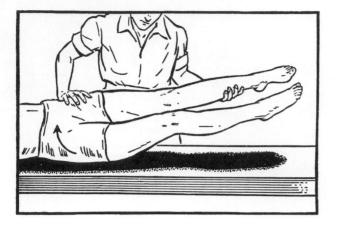

HIP ADDUCTION

Note: Patient may attempt to use hip flexors, medially rotating leg and tipping pelvis backward toward table. Sidelying position must be maintained.

PRIME MOVERS

MUSCLE	ORIGIN	INSERTION
Adductor longus N: Obturator (L3, 4)	a. Anterior surface of pubis at angle of junction of crest with symphysis	a. Middle portion of medial lip of linea aspera
Pectineus N: Femoral (L2, 3, 4)	a. Pectineal line and area just anterior to it, between iliopectineal eminence and tubercle of pubis	a. Line between lesser trochanter and linea aspera
Gracilis N: Obturator (L3, 4)	a. Inferior half of symphysis pubis b. Superior half of pubic arch	a. Medial surface of tibia distal to condyle

HIP LATERAL ROTATION

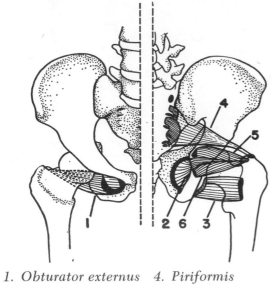

1. *Obturator externus*
2. *Obturator internus*
3. *Quadratus femoris*
4. *Piriformis*
5. *Gemellus superior*
6. *Gemellus inferior*

Range of Motion:

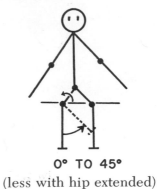

0° TO 45°

(less with hip extended)

Factors Limiting Motion:

1. Tension of lateral band of iliofemoral ligament
2. Tension of hip medial rotator muscles

Fixation:

Weight of trunk

PRIME MOVERS

MUSCLE	ORIGIN	INSERTION
Obturator externus N: Obturator (L3, 4)	a. Medial side of bony margin of obturator foramen b. Medial two thirds of outer surface of obturator membrane c. Rami of pubis d. Ramus of ischium	a. Posterior surface of neck of femur to trochanteric fossa of femur
Obturator internus N: (L5, 2, 3)	a. Greater part of inner margin of obturator foramen b. Pelvic surface of superior part of greater sciatic foramen and inferior and anterior obturator foramen	a. Through lesser sciatic notch to anterior part of medial surface of greater trochanter proximal to trochanteric fossa
Quadratus femoris N: (L5, S1)	a. Proximal portion of external border of ischial tuberosity	a. Proximal part of linea quadrata of femur
Piriformis N: (S1, 2)	a. Anterior surface of sacrum between first and fourth anterior sacral foramina a. Margin of greater sciatic foramen and anterior surface of sacrotuberous ligament	a. Through greater sciatic foramen to superior border of greater trochanter of femur

(Continued on page 52)

HIP LATERAL ROTATION

NORMAL AND GOOD

Sitting with legs over edge of table.
Use counterpressure above knee to prevent abduction and flexion of hip. Patient grasps edge of table to stabilize pelvis.
Patient laterally rotates thigh.
Resistance is given above ankle joint.

FAIR

Sitting with legs over edge of table.
Use counterpressure above knee.
Patient laterally rotates thigh through range of motion with stabilization of pelvis by patient.

POOR

Backlying with leg in internal rotation.
Stabilize pelvis.
Patient laterally rotates leg through range of motion.

TRACE AND ZERO

Presence of contraction in lateral rotators may be determined by deep palpation behind greater trochanter.

★ ★ ★ ★ ★

Note: Resistance should be given slowly and carefully in tests for rotation of the hip and shoulder. Use of the long lever arm can cause injury to joint structures if not controlled.

PRIME MOVERS

MUSCLE	ORIGIN	INSERTION
Gemellus superior N: (L5, S1, 2, 3)	a. Outer surface of ischial spine	a. Upper margin of tendon of Obturator internus and with it to medial surface of greater trochanter
Gemellus inferior N: (L5, S1)	a. Superior part of ischial tuberosity	a. Lower margin of tendon of Obturator internus and with it to medial surface of greater trochanter
Gluteus maximus (Illustrated on p. 38) N: Inferior gluteal (L5, S1, 2)	a. From posterior gluteal line and lateral lip of iliac crest superior and dorsal to line b. Posterior surface of lower part of sacrum and side of coccyx c. Posterior surface of sacrotuberous ligament	a. Iliotibial band of fascia lata over greater trochanter b. Gluteal tuberosity

Accessory Muscles

Sartorius   Biceps femoris (long head)

FOR NOTES:

HIP MEDIAL ROTATION

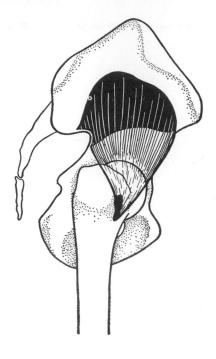

Gluteus minimus

Range of Motion:

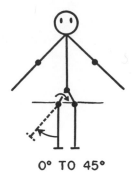

0° TO 45°

(less with hip extended)

Factors Limiting Motion:

1. When hip is extended, tension of iliofemoral ligament
2. When hip is flexed, tension of ischiocapsular ligament
3. Tension of hip lateral rotator muscles

Fixation:

Weight of trunk

PRIME MOVERS

Muscle	Origin	Insertion
Gluteus minimus N: Superior gluteal (L4, 5, S1)	a. Outer surface of ilium between anterior and inferior gluteal lines b. Margin of greater sciatic notch	a. Anterior aspect of greater trochanter of femur b. Expansion to capsule of hip joint
Tensor fasciae latae (Not illustrated) N: Superior gluteal (L4, 5, S1)	a. Anterior part of outer lip of iliac crest b. Outer surface of anterior superior iliac spine	a. Between 2 layers of iliotibial band at juncture of middle and proximal thirds (iliotibial band inserts on lateral condyle of tibia)

Accessory Muscles

Gluteus medius (anterior fibers)
Semitendinosus
Semimembranosus

HIP MEDIAL ROTATION

NORMAL AND GOOD

Sitting with legs over edge of table.
Use counterpressure above knee to prevent
   adduction of the hip. (Patient grasps edge
   of table to stabilize pelvis.)
Patient medially rotates thigh.
Resistance is given above ankle joint.

FAIR

Sitting with legs over table.
Use counterpressure above knee.
Patient medially rotates thigh through range
   of motion with stabilization of pelvis.

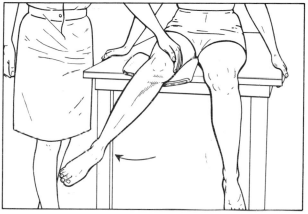

POOR

Backlying with leg in lateral rotation.
Stabilize pelvis.
Patient medially rotates leg through range
   of motion.

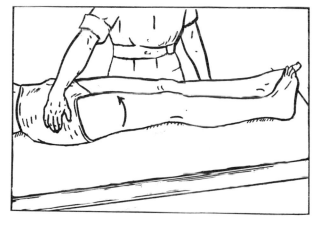

TRACE AND ZERO

Tensor fasciae latae may be palpated near
   its origin posterior and distal to anterior
   superior spine of ilium. Gluteus minimus
   fibers lie beneath Gluteus medius and
   Tensor fasciae latae.

★ ★ ★ ★ ★

Note: If patient lifts pelvis on side being
   tested to assist in medial rotation, pelvis
   should be stabilized. (Not illustrated.)

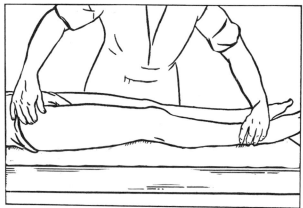

KNEE FLEXION

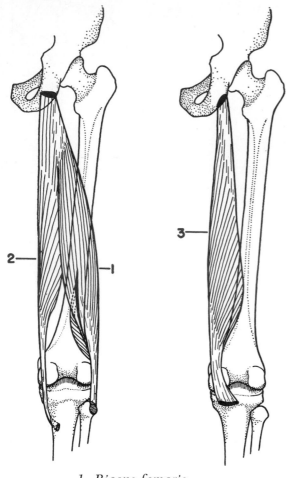

1. *Biceps femoris*
2. *Semitendinosus*
3. *Semimembranosus*

Range of Motion:

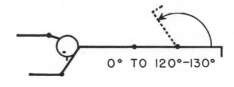

0° TO 120°-130°

Factors Limiting Motion:

   1. Tension of knee extensor muscles, particularly Rectus femoris if hip is extended
   2. Contact of calf with posterior thigh

Fixation:

   1. Contraction of Iliocostalis lumborum and Quadratus lumborum muscles
   2. Weight of thigh and pelvis

PRIME MOVERS

MUSCLE	ORIGIN	INSERTION
Biceps femoris (long head) N: Sciatic (tibial) (S1, 2, 3)	a. Distal and medial impression on ischial tuberosity	a. Lateral side of head of fibula b. Lateral condyle of tibia
Biceps femoris (short head) N: Sciatic (common peroneal) (L4, 5, S1, 2)	a. Whole length of lateral lip of linea aspera and proximal part of lateral supracondylar line of femur	
Semitendinosus N: Sciatic (tibial) (L4, 5, S1, 2, 3)	a. Distal and medial impression on ischial tuberosity	a. Anteromedial surface of tibia at proximal end of shaft

(Continued on page 58)

KNEE FLEXION

NORMAL AND GOOD
(Biceps femoris)

Facelying with legs straight.

Stabilize pelvis.

Patient flexes knee. Grasping above ankle, laterally rotate leg (muscle is placed in better alignment), and resist flexion to test Biceps femoris.

NORMAL AND GOOD
(Semitendinosus and Semimembranosus)

Facelying with legs straight.

Stabilize pelvis.

Patient flexes knee. Grasping proximal to the ankle, medially rotate leg and resist flexion to test Semimembranosus and Semitendinosus.

FAIR

Facelying with legs straight.

Stabilize thigh medially and laterally without pressure over the muscle group being tested.

Patient flexes knee through range of motion. (If Gastrocnemius is weak, knee may be flexed to 10 degrees for starting position.) If Biceps femoris is stronger, lower leg will laterally rotate, if Semitendinosus and Semimembranosus are stronger, lower leg will medially rotate during flexion.

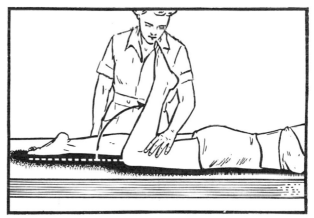

POOR

Sidelying with legs straight and upper leg supported.

Stabilize thigh.

Patient flexes knee through range of motion.

Uneven muscular pull will cause rotation of lower leg as above.

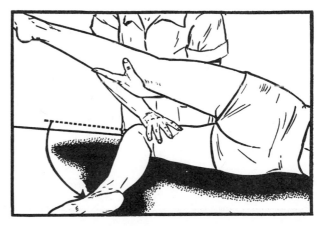

KNEE FLEXION

TRACE AND ZERO

Facelying with knee partially flexed and lower leg supported.

Patient attempts to flex knee. Tendons of knee flexor muscles may be palpated on back of thigh near knee joint.

Note: Patient may flex hip in order to start movement with knee partially flexed.

Notes: The Sartorius may be substituted, which causes flexion and lateral rotation of hip. Knee flexion in this position is less difficult, since lower leg is not raised vertically against gravity. (Not illustrated.)

Strong plantar flexion of the foot should not be allowed in order to prevent substitution by the Gastrocnemius.

PRIME MOVERS

MUSCLE	ORIGIN	INSERTION
Semimembranosus N: Sciatic (tibial) (L4, 5, S1, 2, 3)	a. Proximal and outer impression on ischial tuberosity	a. Horizontal groove on posteromedial aspect of medial condyle of tibia b. Tendon of insertion gives off fibrous expansion into posterior aspect of lateral femoral condyle

Accessory Muscles

Popliteus   Gracilis

Sartorius   Gastrocnemius

FOR NOTES:

KNEE EXTENSION

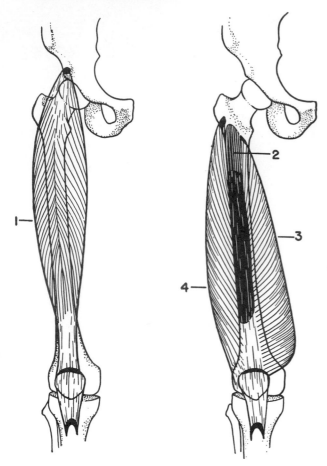

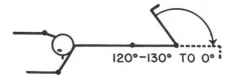

Range of Motion:

120°-130° TO 0°

Factors Limiting Motion:

1. Tension of oblique popliteal, cruciate and collateral ligaments of knee joint
2. Tension of knee flexor muscles

Fixation:

1. Contraction of anterior abdominal muscles to fix origin of Rectus femoris
2. Weight of thigh and pelvis

Quadriceps femoris
1. *Rectus femoris*
2. *Vastus intermedius*
3. *Vastus medialis*
4. *Vastus lateralis*

PRIME MOVERS

MUSCLE	ORIGIN	INSERTION
Quadriceps femoris N: Femoral (L2, 3, 4) *Rectus femoris*	a. Anterior inferior spine of ilium (straight head) b. Groove just above brim of acetabulum (reflected head)	a. Base of patella
Vastus intermedius	a. Anterior and lateral surfaces of proximal two thirds of femoral shaft	a. Forms deep part of Quadriceps femoris tendon, inserting into base of patella
Vastus medialis	a. Distal half of intertrochanteric line b. Medial lip of linea aspera and proximal part of medial supracondylar ridge	a. Medial border of patella and Quadriceps femoris tendon
Vastus lateralis	a. Proximal portion of intertrochanteric line b. Anterior and inferior borders of greater trochanter c. Proximal half of lateral lip of linea aspera	a. Lateral border of patella, forming part of Quadriceps femoris tendon

KNEE EXTENSION

NORMAL AND GOOD

Sitting with legs over edge of table.

Stabilize pelvis without pressure over Rectus femoris at origin.

Patient extends knee through range of motion without terminal locking. (Resistance to a locked knee can be injurious to the joint and is not a valid indicator of the strength of the extensors as a co-contraction of other muscles around the knee is required for the locking action.)

Resistance is given above ankle joint. (Pad should be used under knee.)

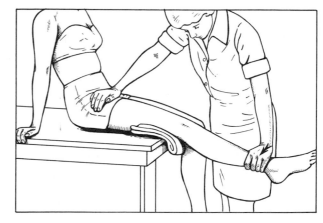

FAIR

Sitting with legs over edge of table.

Stabilize pelvis.

Patient extends knee through range of motion without medial or lateral rotation at the hip (rotation allows extension at an angle, not in a vertical line against gravity).

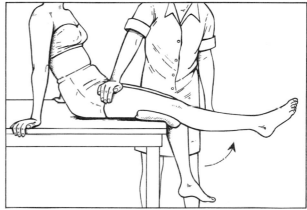

POOR

Sidelying with upper leg supported. Leg to be tested is flexed.

Stabilize thigh above knee joint. (Avoid pressure over Quadriceps femoris.)

Patient extends knee through range of motion.

TRACE AND ZERO

Backlying with knee flexed and supported.

Patient attempts to extend knee.

Contraction of Quadriceps femoris is determined by palpation of tendon between patella and tuberosity of tibia and fibers of muscle. (Latter not illustrated.)

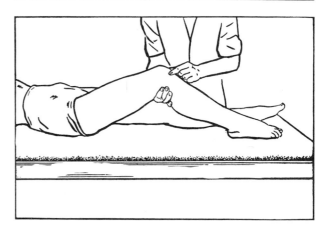

ANKLE PLANTAR FLEXION

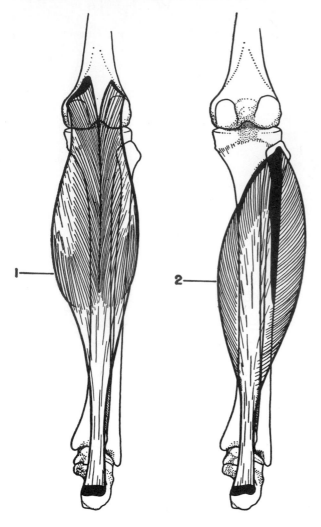

Range of Motion:

0° TO 40°–45°

Factors Limiting Motion:

    1. Tension on anterior talofibular ligament and anterior fibers of deltoid ligament
    2. Tension of dorsiflexor muscles of ankle
    3. Contact of posterior portion of talus with tibia

Fixation:

    Weight of thigh

1. Gastrocnemius
2. Soleus

PRIME MOVERS

MUSCLE	ORIGIN	INSERTION
Gastrocnemius N: Tibial (S1, 2)	*Medial head:* a. Depression on proximal and posterior parts of medial condyle of femur and adjacent area *Lateral head:* a. Impression on side of lateral condyle and posterior surface of femur just proximal to it	a. Tendo calcaneus, which inserts into middle part of posterior surface of calcaneus
Soleus N: Tibial (S1, 2)	a. Posterior surface of head of fibula b. Proximal third of posterior surface of body of fibula c. Popliteal line and middle third of medial border of tibia	a. Tendo calcaneus

	Accessory Muscles	
Tibialis posterior Peroneus longus	Peroneus brevis Flexor hallucis longus	Flexor digitorum longus Plantaris

ANKLE PLANTAR FLEXION

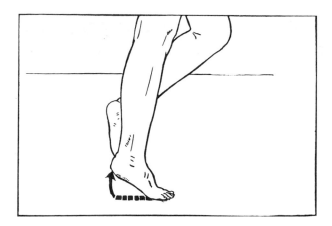

NORMAL AND GOOD

Standing on leg to be tested. Knee straight.
For *Normal* grade, patient raises heel from floor through range of motion of plantar flexion. Patient can complete motion four or five times in good form. (The Tibialis posterior and the Peroneus longus and brevis muscles must be normal or good to stabilize the forefoot and assist with motion by providing counterpressure against the floor.)

A *Good* grade is given if patient has difficulty in completing range of motion or fatigues easily.

FAIR

Standing on leg to be tested. Knee straight.
Patient plantar flexes foot sufficiently to clear heel from floor. (Not illustrated.)

POOR

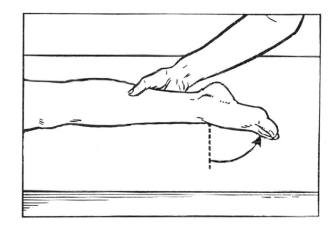

Sidelying with leg to be tested resting on lateral surface; knee extended and foot in midposition.
Stabilize lower leg.
Patient plantar flexes foot through range of motion.

TRACE AND ZERO

Contraction of Gastrocnemius and Soleus are determined by palpation of tendon above calcaneus and muscle fibers on posterior aspect of leg. (Not illustrated.)

NONWEIGHT-BEARING TESTS

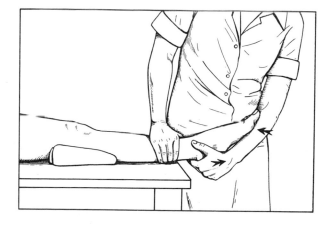

Backlying with pad under knee to prevent hyperextension.
Stabilize leg proximal to ankle.
Patient plantar flexes foot.
Resistance is given by grasping around the calcaneous and exerting pressure against the pull of the plantar flexors. Counterpressure may be given with the forearm against the sole of the foot if the accessory muscles that stabilize the forefoot are functioning.
(Grades of *Normal*, *Good* and *Fair* may be based on amount of resistance given; however, they should be recorded with a question mark, since only weight-bearing tests give accurate grading of strength of Gastronemius and Soleus.)

FOOT DORSIFLEXION AND INVERSION

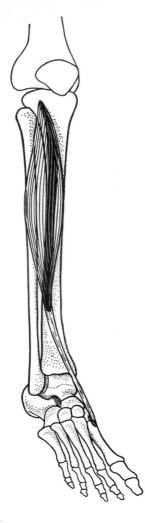

Tibialis anterior

Range of Motion:

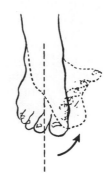

Factors Limiting Motion:

1. Tension of lateral tarsal ligaments
2. Tension of Peroneus longus and Peroneus brevis muscles
3. Contact of tarsal bones medially

Fixation:

Weight of leg

PRIME MOVER

Muscle	Origin	Insertion
Tibialis anterior N: Deep peroneal (L4, 5, S1)	a. Lateral condyle and proximal two thirds of anterolateral surface of tibial body b. Interosseous membrane	a. Medial and plantar surfaces of first cuneiform bone b. Base of first metatarsal bone

FOOT DORSIFLEXION AND INVERSION

NORMAL AND GOOD

Sitting with legs over edge of table.
Stabilize lower leg.
Patient dorsiflexes and inverts foot, keeping
toes relaxed. Resistance is given on
medial dorsal aspect of foot.
Note: Patient should be cautioned to keep
big toe relaxed or flexed to avoid substitution by Extensor hallucis longus.

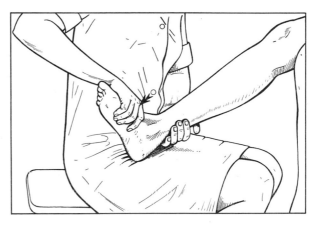

FAIR AND POOR

Sitting with legs over edge of table.
Stabilize lower leg.
Patient inverts and dorsiflexes foot through
full range of motion for fair grade or
through partial range for poor.

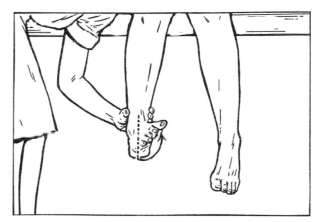

TRACE AND ZERO

Tendon of Tibialis anterior may be palpated
on medial volar aspect of ankle.

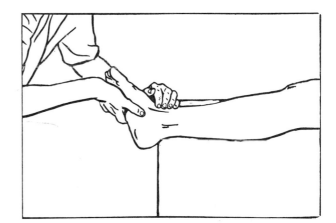

FOOT INVERSION

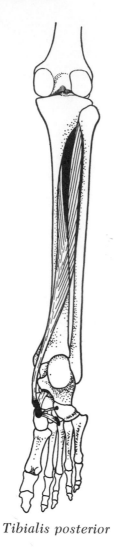

Tibialis posterior

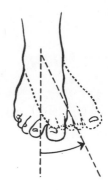

Factors Limiting Motion:

1. Tension of lateral tarsal ligaments
2. Tension of peroneal muscle group
3. Contact of tarsal bones medially

Fixation:

Weight of leg

PRIME MOVER

MUSCLE	ORIGIN	INSERTION
Tibialis posterior N: Tibial (L5, S1)	a. Proximal two thirds of medial surface of fibula b. Lateral portion of posterior surface of tibial body between beginning of popliteal line proximally and junction of middle and lower thirds of body distally c. Interosseous membrane	a. Tuberosity of navicular bone b. Fibrous expansions to sustentaculum tali of calcaneus, to 3 cuneiforms, cuboid, and bases of second, third and fourth metatarsal bones

Accessory Muscles

Flexor digitorum longus
Flexor hallucis longus
Gastrocnemius (medial head)

FOOT INVERSION

NORMAL AND GOOD

Sidelying with foot plantar flexed.
Stabilize lower leg. Avoid pressure over Tibialis posterior muscle.
Patient moves foot through range of motion in inversion.
Resistance is given on medial border of forefoot.

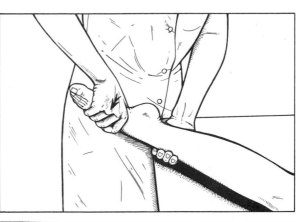

FAIR

Sidelying with foot plantar flexed and resting on lateral border.
Stabilize lower leg. Avoid pressure over Tibialis posterior muscle.
Patient raises foot through range of motion in inversion.

POOR

Backlying with foot in plantar flexion over end of table.
Stabilize lower leg.
Patient inverts foot through range of motion.

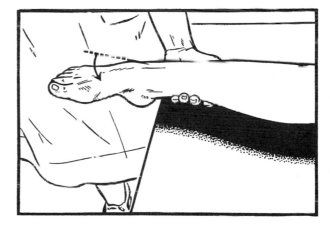

TRACE AND ZERO

Tendon of Tibialis posterior may be found between medial malleolus and navicular bone. It is also palpable above malleolus.

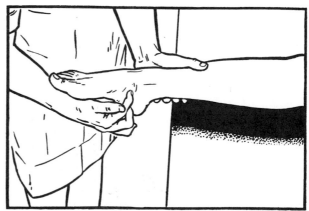

FOOT EVERSION

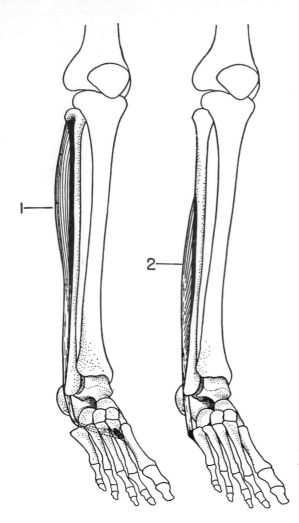

Factors Limiting Motion:

   1. Tension of medial tarsal ligaments
   2. Tension of Tibialis anterior and Tibialis posterior muscles
   3. Contact of tarsal bones laterally

Fixation:

   Weight of leg

1. *Peroneus longus*
2. *Peroneus brevis*

PRIME MOVERS

MUSCLE	ORIGIN	INSERTION
Peroneus longus N: Superficial peroneal (L4, 5, S1)	a. Head and proximal two thirds of lateral surface of body of fibula b. Occasionally a few fibers from lateral condyle of tibia	a. Tendon passes behind lateral malleolus, obliquely forward to groove on inferior surface of cuboid and sole of foot to lateral side of base of first metatarsal and medial cuneiform bones
Peroneus brevis N: Superficial peroneal (L4, 5, S1)	a. Distal two thirds of lateral surface of body of fibula	a. Behind lateral malleolus to tuberosity of base of fifth metatarsal on lateral side of bone

Accessory Muscles

Extensor digitorum longus	Peroneus tertius

FOOT EVERSION

NORMAL AND GOOD

Sidelying with foot plantar flexed.

Stabilize lower leg.

Patient everts foot and depresses head of first metatarsal. To test Peroneus brevis, resistance is given on lateral border of foot. To test Peroneus longus, resistance is given against plantar surface of first metatarsal head. They may be tested together using a derotating motion as illustrated. Extensor digitorum longus should remain relaxed.

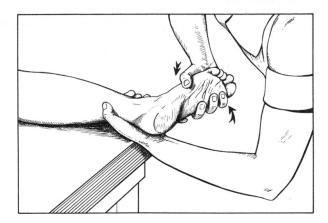

FAIR

Sidelying with foot plantar flexed and resting on medial border.

Stabilize lower leg.

Patient everts foot through range of motion and depresses first metatarsal.

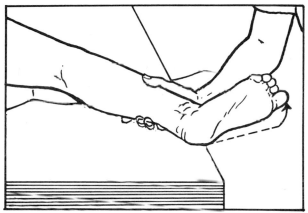

POOR

Backlying with foot in plantar flexion over end of table.

Stabilize lower leg.

Patient everts foot through range of motion with depression of first metatarsal.

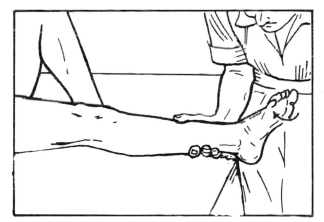

TRACE AND ZERO

Tendon of Peroneus brevis may be found proximal to base of fifth metatarsal on lateral border of foot.

Contraction in Peroneus longus may be determined by light upward pressure under head of first metatarsal.

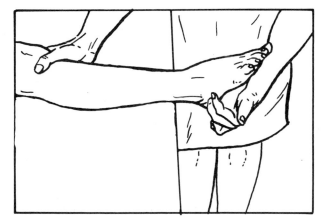

FLEXION OF METATARSOPHALANGEAL JOINTS OF TOES

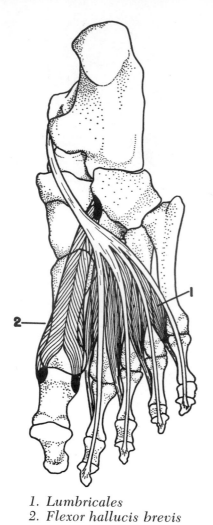

Range of Motion:

0° TO 25°–35°

M.P.

Factors Limiting Motion:

1. Tension of extensor muscle tendons of toes
2. Contact of soft parts

Fixation:

Weight of leg and foot

1. Lumbricales
2. Flexor hallucis brevis

Flexion of Metatarsophalangeal Joints of Lateral Four Toes

PRIME MOVER

MUSCLE	ORIGIN	INSERTION
Lumbricales First lumbricalis: N: Medial plantar (L4, 5) Second, third and fourth lumbricales: N: Lateral plantar (S1, 2)	a. Tendons of Flexor digitorum longus, as far back as their divisions	a. Pass distally around medial sides of lateral 4 toes to insert into expansions of tendons of Extensor digitorum longus on dorsal surfaces of proximal phalanges

Accessory Muscles

Interossei dorsales and plantares
Flexor digiti quinti brevis
Flexor digitorum longus
Flexor digitorum brevis

(Continued on page 71.)

FLEXION OF METATARSOPHALANGEAL JOINTS OF TOES

FLEXION OF METATARSOPHALANGEAL JOINTS OF LATERAL FOUR TOES (LUMBRICALES)

Backlying.

Stabilize metatarsals.

Patient flexes lateral four toes. Resistance is given beneath proximal row of phalanges for normal and good.

Note: Individual testing of the four lateral toes is often desirable in all tests as they vary in strength.

FLEXION OF METATARSOPHALANGEAL JOINT OF HALLUX (FLEXOR HALLUCIS BREVIS)

Backlying.

Stabilize first metatarsal.

Patient flexes hallux. Resistance is given beneath proximal phalanx for normal and good.

Note: Grades below the level of normal and good may be difficult to determine as joint motion is often limited and the muscles and tendons cannot be palpated. If range appears normal, a grade of fair may be given for completion of full motion, a poor for partial range.

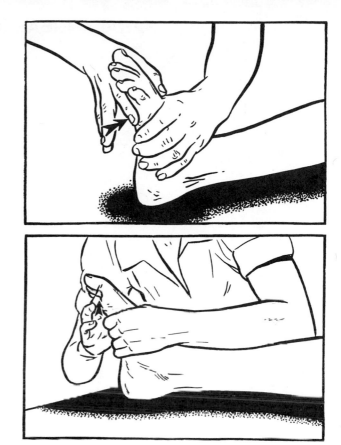

Flexion of Metatarsophalangeal Joint of Hallux

PRIME MOVERS

MUSCLE	ORIGIN	INSERTION
Flexor hallucis brevis N: Medial plantar (L4, 5, S1)	a. Medial part of plantar surface of cuboid b. Contiguous portion of lateral cuneiform bone	a. By 2 tendons into medial and lateral sides of base of proximal phalanx of hallux (sesamoid bone in each tendon)

Accessory Muscle
Flexor hallucis longus

FLEXION OF INTERPHALANGEAL JOINTS OF TOES

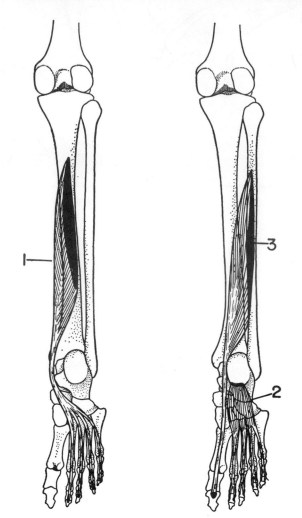

Range of Motion:

0° TO 50°–90°

Factors Limiting Motion:

1. Tension of extensor muscle tendons (dorsal ligaments) of toes
2. Contact of soft parts of phalanges

Fixation:

1. Stabilization of foot by invertor and evertor muscle groups
2. Weight of leg and foot
3. Synergistic action of Tibialis anterior to prevent plantar flexion of ankle

1. *Flexor digitorum longus*
2. *Flexor digitorum brevis*
3. *Flexor hallucis longus*

Flexion of Distal Interphalangeal Joints of Lateral Four Toes

PRIME MOVER

MUSCLE	ORIGIN	INSERTION
Flexor digitorum longus N: Tibial (L5, S1)	a. Posterior surface of body of tibia distal to the popliteal line to within 7 or 8 cm. of its distal end	a. Bases of distal phalanges of lateral 4 toes

Flexion of Proximal Interphalangeal Joints of Lateral Four Toes

PRIME MOVER

MUSCLE	ORIGIN	INSERTION
Flexor digitorum brevis N: Medial plantar (L4, 5)	a. Medial process of tuberosity of calcaneus	a. By 2 slips into sides of second phalanges of lateral 4 toes

FLEXION OF INTERPHALANGEAL JOINTS OF TOES

FLEXION OF DISTAL INTERPHALANGEAL JOINTS OF LATERAL FOUR TOES (FLEXOR DIGITORUM LONGUS)

Backlying.
Stabilize middle row of phalanges of lateral four toes.
Patient flexes toes. Resistance is given beneath third row of phalanges of lateral four toes for normal and good.

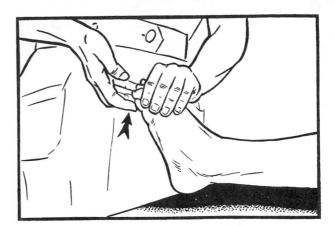

FLEXION OF PROXIMAL INTERPHALANGEAL JOINTS OF LATERAL FOUR TOES (FLEXOR DIGITORUM BREVIS)

Backlying.
Stabilize proximal row of phalanges of lateral four toes.
Patient flexes toes. Resistance is given beneath second row of phalanges of lateral four toes for normal and good.

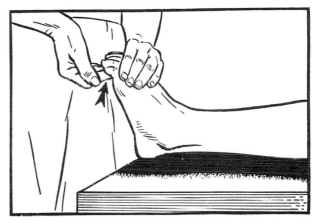

FLEXION OF INTERPHALANGEAL JOINT OF HALLUX (FLEXOR HALLUCIS LONGUS)

Backlying.
Stabilize proximal phalanx of hallux.
Patient flexes toe.
Resistance is given beneath second phalanx for normal and good.
(For *Trace* and *Zero* grades, tendon of Flexor hallucis longus may be found on plantar surface of proximal phalanx.)

Note: Joint motion may be limited. If range appears normal, a grade of fair may be given for completion of motion, a poor for partial range.

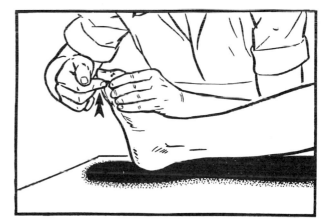

Flexion of Interphalangeal Joint of Hallux

PRIME MOVER

MUSCLE	ORIGIN	INSERTION
Flexor hallucis longus N: Tibial (L5, S1, 2)	a. Inferior two thirds of posterior surface of body of fibula a. Lower part of interosseous membrane	a. Base of terminal phalanx of great toe

EXTENSION OF METATARSOPHALANGEAL JOINTS OF TOES AND INTERPHALANGEAL JOINT OF HALLUX

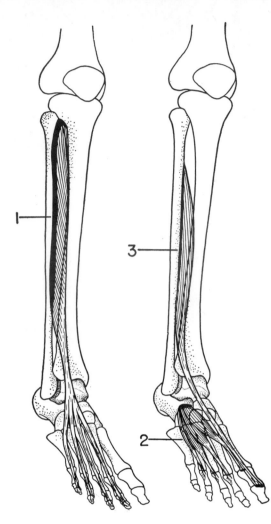

Range of Motion:

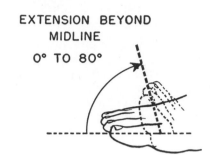

EXTENSION BEYOND MIDLINE

0° TO 80°

Factors Limiting Motion:

Tension of plantar and collateral ligaments of toe joints

Fixation:

1. Weight of leg and foot
2. Synergistic action of Gastrocnemius and Soleus muscles to prevent dorsiflexion of ankle (with contraction of long toe extensors)

1. *Extensor digitorum longus*
2. *Extensor digitorum brevis*
3. *Extensor hallucis longus*

Extension of Metatarsophalangeal Joints of Lateral Four Toes and Hallux

PRIME MOVERS

MUSCLE	ORIGIN	INSERTION
Extensor digitorum longus N: Deep peroneal (L4, 5, S1)	a. Lateral condyle of tibia b. Proximal three fourths of anterior surface of fibula	a. Extensor expansions into second and third phalanges of lateral 4 toes
Extensor digitorum brevis N: Deep peroneal (L5, S1)	a. Distal and lateral surfaces of calcaneus in front of groove for Peroneus brevis muscle	a. Medial division into dorsal surface of proximal phalanx of hallux at base (sometimes called Extensor hallucis brevis) b. Three lateral divisions into tendon of Extensor digitorum longus of second, third and fourth toes

EXTENSION OF METATARSOPHALANGEAL JOINTS OF TOES AND INTERPHALANGEAL JOINT OF HALLUX

EXTENSION OF METATARSOPHALANGEAL JOINTS OF LATERAL FOUR TOES (EXTENSOR DIGITORUM LONGUS AND EXTENSOR DIGITORUM BREVIS)

Backlying.

Stabilize metatarsal area.

Patient extends lateral four toes. Resistance is given on proximal phalanges of toes for normal and good.

(For *Trace* and *Zero* grades, tendons of Extensor digitorum longus may be palpated on dorsal surface of metatarsals and fibers of Extensor digitorum brevis on lateral side of dorsum of foot anterior to malleolus.)

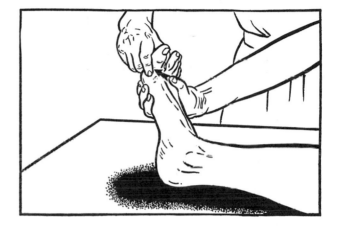

EXTENSION OF METATARSOPHALANGEAL JOINT OF HALLUX (MEDIAL DIVISION OF EXTENSOR DIGITORUM BREVIS)

Backlying.

Stabilize first metatarsal.

Patient extends metatarsophalangeal joint of hallux. Resistance is given over proximal phalanx for normal and good.

EXTENSION OF INTERPHALANGEAL JOINT OF HALLUX (EXTENSOR HALLUCIS LONGUS)

Backlying.

Stabilize proximal phalanx of hallux.

Patient extends distal joint of hallux. Resistance is given on dorsal surface for normal and good.

(For *Trace* and *Zero* grades, tendon of Extensor hallucis longus may be palpated over the dorsal surface of first metatarsophalangeal joint and on a diagonal line across dorsum of foot to middle of anterior aspect of ankle.)

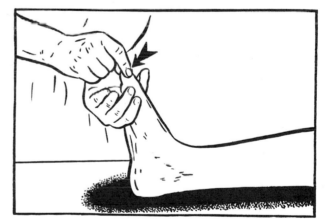

See note on page 73 for statement concerning fair and poor grading.

Extension of Interphalangeal Joint of Hallux

PRIME MOVER

MUSCLE	ORIGIN	INSERTION
Extensor hallucis longus N: Deep peroneal (L4, 5, S1)	a. Middle 2 fourths of anterior surface of fibula	a. Base of distal phalanx of hallux

SCAPULAR ABDUCTION AND UPWARD ROTATION:

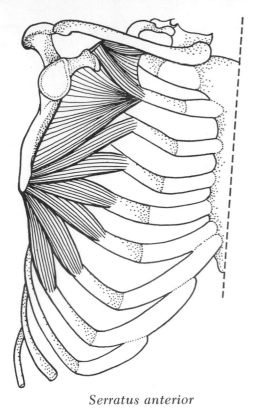

Serratus anterior

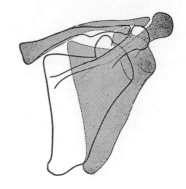

Range of Motion:

Factors Limiting Motion:

1. Tension of trapezoid ligament (limits forward rotation of scapula upon clavicle)
2. Tension of Trapezius and Rhomboid major and minor muscles

Fixation:

1. In strong scapular abduction, pull of external Obliquus externus abdominus on same side
2. Weight of thorax

*Drawn from x-ray views of subject at beginning and end of movement in test for normal muscle. Change in shape of scapula is due to shift in plane during movement.

PRIME MOVER

MUSCLE	ORIGIN	INSERTION
Serratus anterior N: Long thoracic (C5, 6, 7)	a. Digitations from outer surfaces and superior borders of upper 8 or 9 ribs b. Aponeuroses covering intercostal muscles	a. Ventral surface of superior angle b. Ventral surface of vertebral border of scapula. c. Lower 5 or 6 digitations converge and insert on ventral surface of inferior angle of scapula.

SCAPULAR ABDUCTION AND UPWARD ROTATION

NORMAL AND GOOD

Backlying with arm flexed to 90 degrees with slight abduction, and elbow in extension.

Patient moves arm upward by abducting the scapula. Resistance is given by grasping around forearm and elbow. Pressure is downward and inward toward table. Observe scapula for "winging" (movement of vertebral border away from thorax) and substitution by anterior muscles of shoulder.

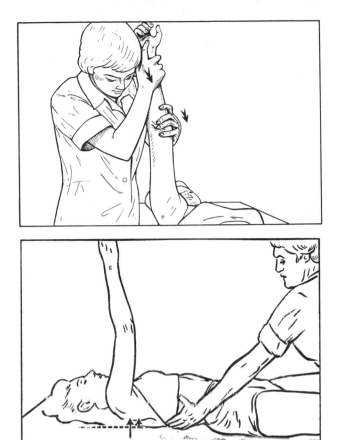

FAIR

Backlying with arm flexed to 90 degrees and scapula resting on table.

Stabilize thorax.

Patient forces arm upward. Scapula should be completely abducted without "winging." (If extensor muscles of elbow are weak, elbow may be flexed or forearm may be supported.)

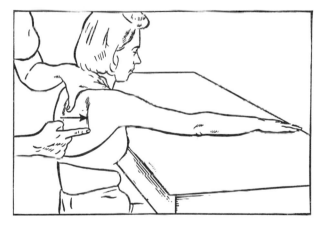

POOR

Sitting with arm flexed to 90 degrees and arm resting on table.

Stabilize thorax.

Patient moves arm forward by abducting scapula

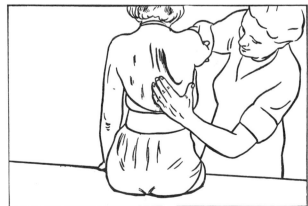

TRACE AND ZERO

Examiner lightly forces arm backward to determine presence of a contraction of Serratus anterior. Scapula should be observed for "winging." Digitations of Serratus anterior may be palpated on outer surface of ribs for a contraction. (Latter not illustrated.)

SCAPULAR ELEVATION

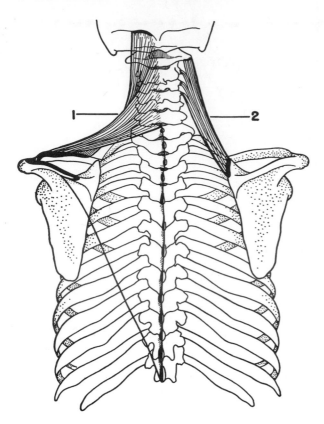

1. *Trapezius (superior fibers)*
2. *Levator scapulae*

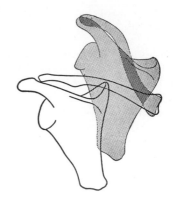

Range of Motion:*

Factors Limiting Motion:

   1. Tension of costoclavicular ligament
   2. Tension of muscles depressing scapula and clavicle: Pectoralis minor, Subclavius, and Trapezius (lower fibers)

Fixation:

   1. Flexor muscles of cervical spine (for tests done in sitting position)
   2. Weight of head (for tests done in face-lying position)

*Drawn from x-ray views of subject at beginning and end of movement in test for normal muscle.

PRIME MOVERS

MUSCLE	ORIGIN	INSERTION
Trapezius (superior fibers) N: Spinal accessory and branches from (C3, 4)	a. External occipital protuberance b. Medial third of superior nuchal line of occipital bone c. Upper part of ligamentum nuchae	a. Posterior border of lateral third of clavicle
Levator scapulae N: (C3, 4) and frequently branch from dorsal scapular	a. Transverse processes of atlas, axis, third and fourth cervical vertebrae	a. Vertebral border of scapula superior to root of spine

Accessory Muscles

Rhomboideus major and minor

SCAPULAR ELEVATION

NORMAL AND GOOD

Sitting with arms at sides.
Patient raises shoulders as high as possible.
   Resistance is given downward on top of
   shoulders.

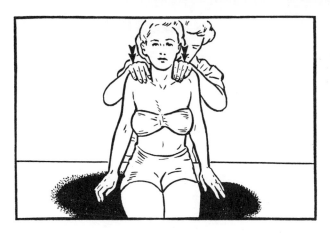

FAIR

Sitting with arms at sides.
Patient elevates shoulders through range of
motion.

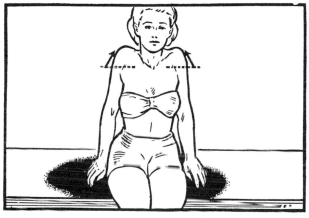

POOR

Facelying with shoulders supported by examiner and forehead resting on table.
Patient moves shoulders toward ears
   through range of motion.

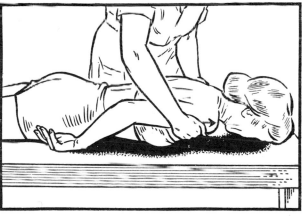

TRACE AND ZERO

Facelying.
Examiner palpates upper fibers of Trapezius
   parallel to cervical vertebrae and near
   their insertion above clavicle. (Latter not
   illustrated.)

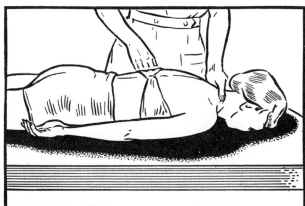

SCAPULAR ADDUCTION

Range of Motion:*

Factors Limiting Motion:

1. Tension of conoid ligament (limits backward rotation of scapula upon clavicle)
2. Tension of Pectoralis major and minor and Serratus anterior muscles
3. Contact of vertebral border of scapula with spinal musculature

Fixation:

Weight of trunk

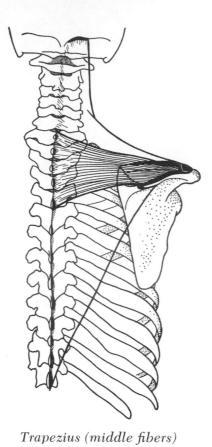

Trapezius (middle fibers)

*Drawn from x-ray views of subject at beginning and end of movement in test for normal muscles.

PRIME MOVERS

MUSCLE	ORIGIN	INSERTION
Trapezius (middle fibers) N: Spinal accessory and branches from (C3, 4)	a. Inferior part of ligamentum nuchae b. Spinous processes of seventh cervical and superior thoracic vertebrae (horizontal fibers of muscle)	a. Medial margin of acromion process of scapula b. Superior lip of posterior border of spine of scapula
Rhomboideus major and *minor* (Illustrated on p. 86) N: Dorsal scapular (C5)	a. Spinous processes of seventh cervical and first 5 thoracic vertebrae	a. Vertebral border of scapula between root of spine and inferior angle

Accessory Muscles

Rhomboideus major and minor
Trapezius (lower and upper fibers)

SCAPULAR ADDUCTION

NORMAL AND GOOD

Facelying with arm abducted to 90 degrees and laterally rotated, elbow flexed to a right angle.

Stabilize thorax.

Patient raises arm in horizontal abduction, motion taking place primarily between the scapula and thorax and not at glenohumeral joint. Scapula is adducted and fixed by middle section of the Trapezius. Resistance is given on lateral angle of scapula. (No pressure is placed on the humerus.)

(If glenohumeral muscles are weak, arm may be placed in a vertical position over edge of table.)

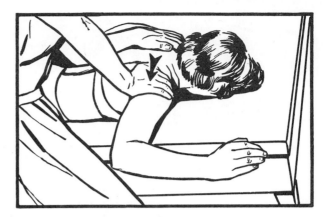

FAIR

Facelying with arm abducted to 90 degrees and laterally rotated, elbow flexed to a right angle.

Stabilize thorax.

Patient raises arm and adducts scapula.

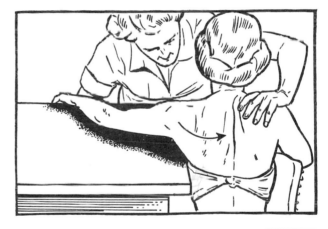

POOR

Sitting with arm resting on table midway between flexion and abduction.

Stabilize thorax.

Patient horizontally abducts arm and adducts scapula.

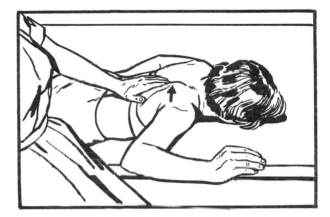

TRACE AND ZERO

Sitting or facelying.

Middle fibers of Trapezius are palpated between root of spine of scapula and vertebral column to determine presence of a contraction.

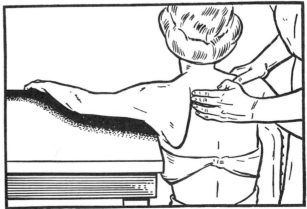

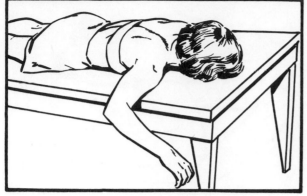

SCAPULAR ADDUCTION

Note: Depression of shoulder toward table with scapular abduction on attempt to perform test motion indicates that posterior fibers of Deltoideus are contracting, but scapula is not fixed or adducted.

Whole muscle exam.
  Sit: hip flex., sart, quad
back - SCM. Abd - post pelvic tilt.
    rot at hip & arm. Poor triceps & biceps
Side. Biceps & triceps. - poor.
Stomach. - all scap muscles. Post
  delt., int & ext rot at sh.
  Poor test for ant delt, pec & latiss
  mid delt - supine. <side   pec & arm   & post delt.
on pillow

FOR NOTES:

SCAPULAR DEPRESSION AND ADDUCTION

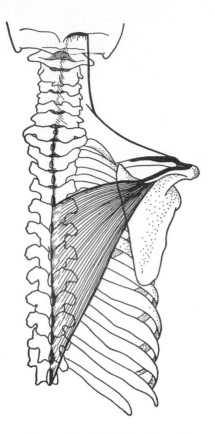

Trapezius (inferior fibers)

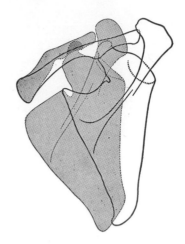

Factors Limiting Motion:

1. Tension of interclavicular ligament and articular disk of sternoclavicular joint
2. Tension of Trapezius (upper fibers), Levator scapulae and Sternocleidomastoideus (clavicular head)

Fixation:

1. Contraction of spinal extensor muscles
2. Weight of trunk

*Drawn from x-ray views of subject at beginning and end of movement in test for normal muscle.

PRIME MOVER

MUSCLE	ORIGIN	INSERTION
Trapezius (inferior fibers) N: Spinal accessory and branches from (C3, 4)	a. Spinous processes of inferior thoracic vertebrae and corresponding supraspinal ligament	a. By aponeurosis gliding over medial end of spine of scapula to tubercle at apex of smooth triangular surface

Accessory Muscles

Trapezius: (middle fibers, adduction)

SCAPULAR DEPRESSION AND ADDUCTION

NORMAL AND GOOD

Facelying with forehead resting on table and arm to be tested extended overhead.

Patient raises arm and fixates scapula strongly with lower part of Trapezius. Resistance is given on lateral angle of scapula in upward and outward direction.

(If shoulder flexion is limited, arm may be placed over edge of table.)

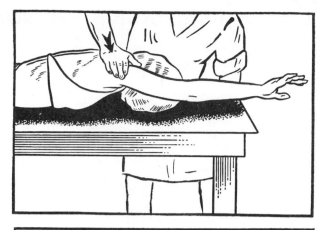

NORMAL AND GOOD
(Alternate)

If Deltoideus is weak, arm is passively raised by examiner.

Patient attempts to assist. Resistance is given on scapula as above.

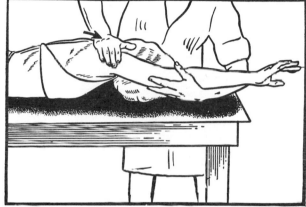

FAIR AND POOR

Facelying with forehead resting on table and arm overhead.

Patient lifts arm from table through full range of motion without upward movement of the scapula or forward sagging of the acromion process for fair grade or through partial range for poor.

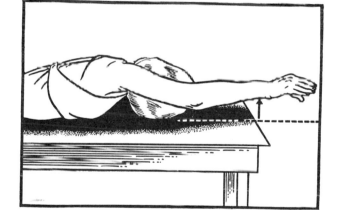

TRACE AND ZERO

Examiner palpates fibers of lower part of Trapezius between last thoracic vertebrae and scapula.

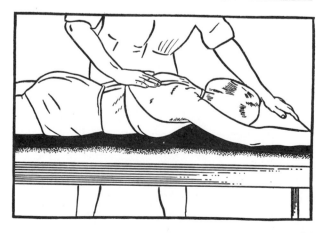

SCAPULAR ADDUCTION AND DOWNWARD ROTATION

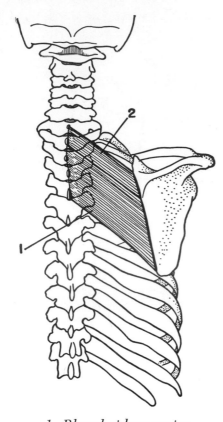

1. Rhomboideus major
2. Rhomboideus minor

Range of Motion:

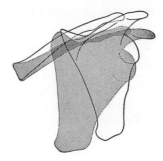

Factors Limiting Motion:

1. Tension of conoid ligament (limits backward rotation of scapula upon clavicle)
2. Tension of Pectoralis major and minor and Serratus anterior muscles
3. Contact of vertebral border of scapula with spinal musculature

Fixation:

Weight of trunk

PRIME MOVERS

MUSCLE	ORIGIN	INSERTION
Rhomboideus major N: Dorsal scapular (C5)	a. Spinous processes of second, third, fourth and fifth thoracic vertebrae	a. Tendinous arch from root of spine of scapula to inferior angle (arch connected to scapula by thin membrane)
Rhomboideus minor N: Dorsal scapular (C5)	a. Inferior part of ligamentum nuchae b. Spinous processes of seventh cervical and first thoracic vertebrae	a. Base of triangular smooth surface at root of spine of scapula

Accessory Muscle

Trapezius (adduction)

*Drawn from x-ray views of subject at beginning and end of movement in test for normal muscles.

SCAPULAR ADDUCTION AND DOWNWARD ROTATION

NORMAL AND GOOD

Facelying with arm medially rotated and adducted across back. Shoulders relaxed.
Patient raises arm and adducts scapula. Resistance is given on vertebral border of scapula in outward and slightly downward direction.

why down?

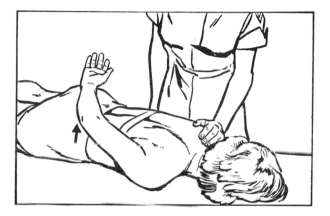

FAIR

Facelying with arm medially rotated and adducted across back and shoulders relaxed.
Patient raises arm and adducts scapula through range of motion.
(If the glenohumeral muscles are weak, slight resistance may be given to the scapula for a fair grade.)

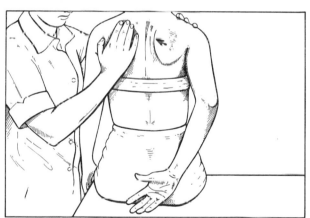

POOR

Sitting with arm medially rotated and adducted behind back. Stabilize trunk with anterior and posterior pressure to prevent flexion and rotation.
Patient adducts scapula through range of motion.

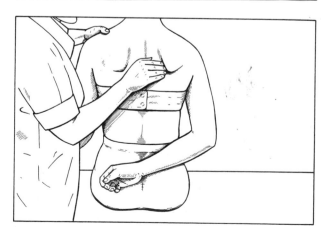

TRACE AND ZERO

Examiner palpates Rhomboid muscles at the angle formed by the vertebral border of the scapula and the lateral fibers of the lower Trapezius.

SHOULDER FLEXION TO 90 DEGREES

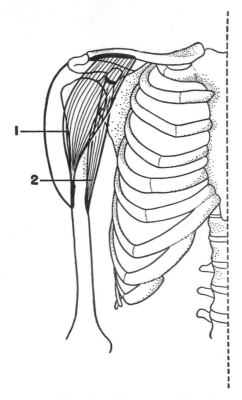

Range of Motion:

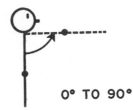

0° TO 90°

Factors Limiting Motion:

None; range of motion incomplete

Fixation:

Contraction of Trapezius and Serratus anterior muscles. (Serratus anterior and upper fibers of Trapezius assist in upward rotation of scapula as well as in fixation.)

scapula abducts.

1. *Deltoideus (anterior fibers)*
2. *Coracobrachialis*

PRIME MOVERS

MUSCLE	ORIGIN	INSERTION
Deltoideus (anterior fibers) N: Axillary (C5, 6)	a. Anterior border and superior surface of lateral third of clavicle	a. Deltoid tubercle on middle of the lateral side of body of humerus
Coracobrachialis N: Musculocutaneous (C6, 7)	a. Apex of coracoid process	a. Medial surface and border of humerus opposite insertion of Deltoideus

Accessory Muscles

Deltoideus (middle fibers)
Pectoralis major (clavicular fibers)
Biceps brachii

SHOULDER FLEXION TO 90 DEGREES

NORMAL AND GOOD

Sitting with arm at side and elbow slightly flexed.

Stabilize scapula.

Patient flexes arm to 90 degrees (palm down to prevent lateral rotation with substitution by the Biceps brachii). Resistance is given above elbow.

(Patient should not be allowed to rotate or horizontally adduct or abduct arm.)

FAIR

Sitting position with arm at side and elbow slightly flexed.

Stabilize scapula.

Patient flexes arm to 90 degrees (palm down).

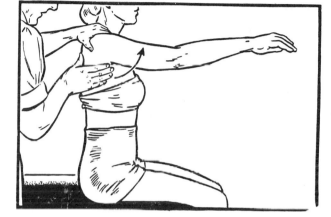

POOR

Sidelying with arm at side resting on smooth board (or arm supported by examiner) and elbow slightly flexed.

Stabilize scapula.

Patient brings arm forward to 90 degrees of flexion.

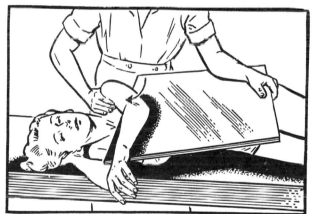

TRACE AND ZERO

Backlying.

Examiner palpates fibers of anterior portion of Deltoideus on anterior aspect of shoulder joint.

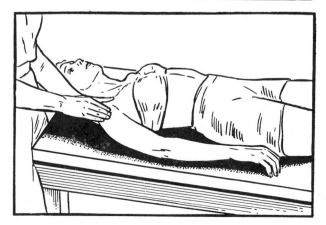

SHOULDER FLEXION TO 90 DEGREES

Note: Patient may laterally rotate upper arm and attempt to flex shoulder with Biceps brachii. Arm should be kept in midposition between medial and lateral rotation.

cont.
tight add & med rot. Both arms
overhead and flat. xInterfer with lower trap
test. Weak serialus. Sh retracted –
then protract.

Oct 31.

Sit. upper trap, - shrug.   Strong R.
Serratus - reach foreward. Bend elbow. R on elbow.
levator or start back in retr. pt pushes you.

Stomach: Med & lat rotators
Latiss - arm hang flex, abd, ~~med~~ ext rot →
ext, add, int~~lat~~ rot. An arc.   Palm up.

arm abd to 90° elbow flex.
med rot -   subscap - arm hangs over
scap. fingers under scap.
keep arm from being add.
lat rot - teres minor right above major.
start in int rot.   Poor. - arm
changing start in int. rot.

Prone: mid, lower trap, rhom.
arm. abd to 90, thumb up - ext rot.
Left arm, add scap. Pinch sh. blades together
arm not pull down.   R- push scap hard
outward or push on arm not as hard.
lower trap. - arm overhead. Pinch
sh blades together - depress & add. R is upt
out toward elev. & abd. Some people can't get
arm up - R on scap.   (hard)
Rhom - arm across back. left arm. Pinch
blades together. Push arm out or push scap
out & up.
Poor - Sit except latiss is on side. Start
arm over head. Around & up in ext. hold
hand. Sit! Mid trap. Arm support, ext rot.
over - arm over
Serratus. Supine. hand under scap
Supine - int or ext rot. if she is beak or
can't turn over.

SHOULDER EXTENSION

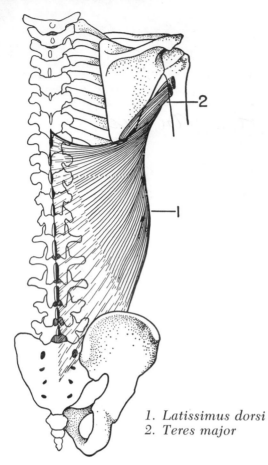

Range of Motion:

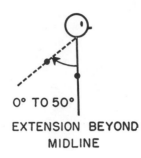

0° TO 50°

EXTENSION BEYOND MIDLINE

Factors Limiting Motion:

    1. Tension of shoulder flexor muscles
    2. Contact of greater tubercle of humerus with acromion posteriorly

Fixation:

    1. Contraction of Rhomboideus major and minor and Trapezius muscles
    2. Weight of trunk

1. Latissimus dorsi
2. Teres major

PRIME MOVERS

MUSCLE	ORIGIN	INSERTION
Latissimus dorsi N: Thoracodorsal (C6, 7, 8)	a. Posterior layer of lumbodorsal fascia by which it is attached to spines of the lower 6 thoracic, lumbar and sacral vertebrae, supraspinal ligament, and to posterior iliac crest b. External lip of iliac crest lateral to Sacrospinalis c. Caudal three or 4 lower ribs d. Usually a few fibers from inferior angle of scapula	a. Bottom of intertubercular groove of humerus
Teres major N: Lowest subscapular (C5, 6)	a. Dorsal surface of inferior angle of scapula	a. Crest below lesser tuberosity of humerus posterior to Latissimus dorsi
Deltoideus (posterior fibers) (Illustrated on p. 96) N: Axillary (C5, 6)	a. Lower lip of posterior border of spine of scapula	a. Deltoid tubercle on middle of the lateral side of body of humerus

Accessory Muscles

Teres minor   Triceps brachii (long head)

SHOULDER EXTENSION

NORMAL AND GOOD

Facelying with arm medially rotated and
adducted (palm up to prevent lateral
rotation).
Stabilize scapula.
Patient extends arm through range of
motion. Resistance is given proximal to
elbow.

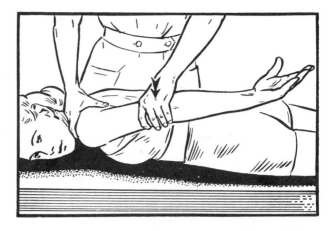

FAIR

Facelying with arm at side.
Stabilize scapula.
Patient extends arm through range of
motion.

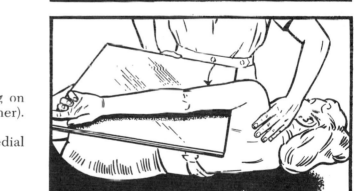

POOR

Sidelying with arm flexed and resting on
smooth board (or supported by examiner).
Stabilize scapula.
Patient extends arm in position of medial
rotation through range of motion.

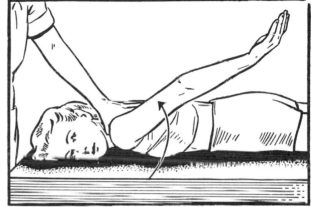

TRACE AND ZERO

Facelying.
Examiner palpates fibers of Teres major on
lower part of axillary border of scapula
(not shown) and fibers of Latissimus dorsi
slightly below.

★ ★ ★ ★ ★

Note: Motion should take place primarily at
glenohumeral joint. Patient should not be
permitted to tip scapula forward in an
effort to complete range of motion. (Not
illustrated.)

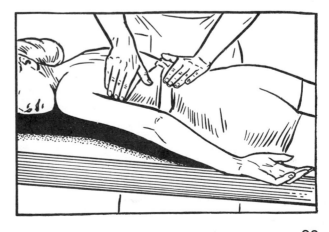

SHOULDER ABDUCTION TO 90 DEGREES

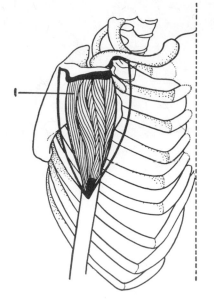

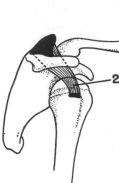

Range of Motion:

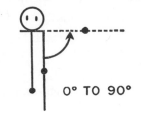

0° TO 90°

Factors Limiting Motion:

None; range of motion incomplete.

Fixation:

Contraction of Trapezius and Serratus anterior muscles. (Serratus anterior and upper fibers of Trapezius assist in upward rotation of scapula as well as in fixation.)

1. *Deltoideus*
2. *Supraspinatus*

PRIME MOVERS

MUSCLE	ORIGIN	INSERTION
Deltoideus (middle fibers) N: Axillary (C5, 6)	a. Lateral margin and superior surface of acromion	a. Deltoid tubercle on middle of the lateral side of body of humerus
Supraspinatus N: Suprascapular (C5)	a. Medial two thirds of supraspinatus fossa	a. Highest of 3 impressions on greater tubercle of humerus

Accessory Muscles

Deltoideus (anterior and posterior fibers)
Serratus anterior (direct action on scapula)

SHOULDER ABDUCTION TO 90 DEGREES

NORMAL AND GOOD

Sitting with arm at side in midposition between medial and lateral rotation. Elbow flexed a few degrees.

Stabilize scapula.

Patient abducts arm to 90 degrees without lateral rotation at shoulder joint (palm down to prevent lateral rotation with substitution by the Biceps brachii). Resistance is given proximal to elbow joint.

SHOULDER ABDUCTION TO 90 DEGREES

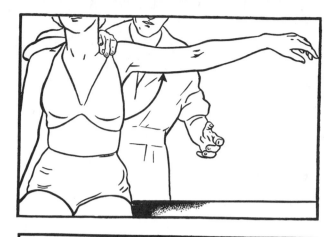

FAIR

Sitting with arm at side in midposition between medial and lateral rotation. Elbow flexed a few degrees.

Stabilize scapula.

Patient abducts arm to 90 degrees without lateral rotation at shoulder joint (palm down).

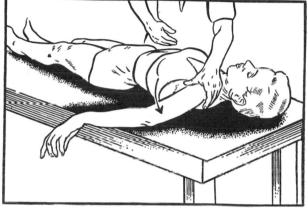

POOR

Backlying with arm at side in midposition between medial and lateral rotation. Elbow slightly flexed.

Stabilize scapula over acromion.

Patient abducts arm to 90 degrees without lateral rotation at shoulder joint.

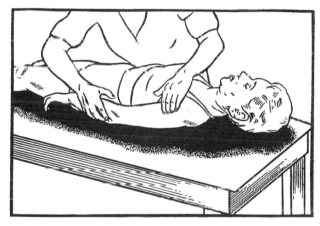

TRACE AND ZERO

Examiner palpates middle section of Deltoideus on lateral surface of upper third of arm.

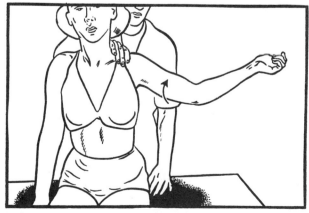

Note: Patient may laterally rotate arm and attempt to substitute Biceps brachii during abduction. Arm should be kept in midposition between medial and lateral rotation.

SHOULDER HORIZONTAL ABDUCTION

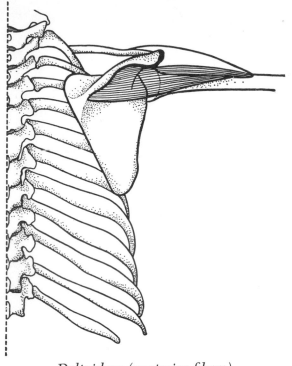

Deltoideus (posterior fibers)

Range of Motion:

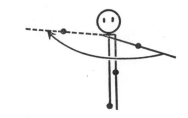

Factors Limiting Motion:

1. Tension of anterior fibers of capsule of glenohumeral joint
2. Tension of Pectoralis major and Deltoideus (anterior fibers)

Fixation:

Contraction of Rhomboideus major and minor and Trapezius (primarily middle and lower fibers)

PRIME MOVER

MUSCLE	ORIGIN	INSERTION
Deltoideus (posterior fibers) N: Axillary (C5, 6)	a. Inferior lip of posterior border of spine of scapula	a. Deltoid tubercle on middle of the lateral side of body of humerus

Accessory Muscles

Infraspinatus
Teres minor

SHOULDER HORIZONTAL ABDUCTION

NORMAL AND GOOD

Facelying with shoulder abducted to 90 degrees, upper arm resting on table and lower arm hanging vertically over edge.
Stabilize scapula.
Patient abducts upper arm through range of motion. Resistance is given proximal to elbow. Motion takes place primarily at glenohumeral joint and not between scapula and thorax.

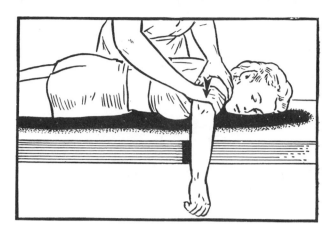

FAIR

Facelying with shoulder abducted to 90 degrees, upper arm resting on table and lower arm hanging vertically over edge.
Stabilize scapula.
Patient abducts upper arm through range of motion.

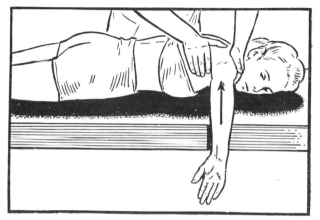

POOR

Sitting with arm supported in a position of 90 degrees of flexion.
Stabilize scapula.
Patient horizontally abducts arm through range of motion.

TRACE AND ZERO

Muscle fibers of posterior portion of Deltoideus are palpated on posterior aspect of shoulder joint.

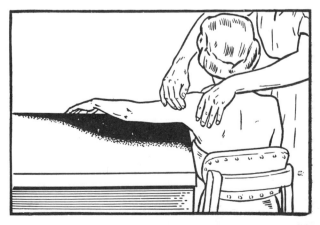

SHOULDER HORIZONTAL ADDUCTION

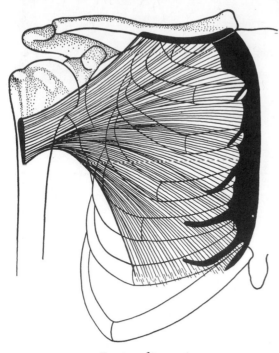

Pectoralis major

Range of Motion:

Factors Limiting Motion:

1. Tension of shoulder extensor muscles
2. Contact of arm with trunk

Fixation:

In forceful horizontal adduction, contraction of Obliquus externus abdominus muscle on same side

PRIME MOVER

MUSCLE	ORIGIN	INSERTION
Pectoralis major N: Medial and lateral pectoral (C5, 6, 7, 8, Th1)	a. Anterior surface of sternal half of clavicle b. Half of breadth of ventral surface of sternum as far as sixth or seventh rib c. Cartilages of first 6 or 7 ribs	a. Crest of greater tubercle of humerus

Accessory Muscle

Deltoideus (anterior fibers)

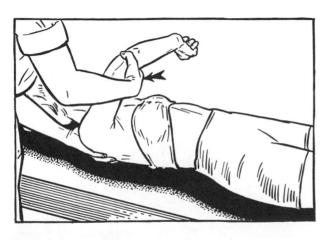

SHOULDER HORIZONTAL ADDUCTION

NORMAL AND GOOD

Backlying with arm abducted to 90 degrees. Patient adducts arm through range of motion. Resistance is given proximal to elbow joint.

SHOULDER HORIZONTAL ADDUCTION

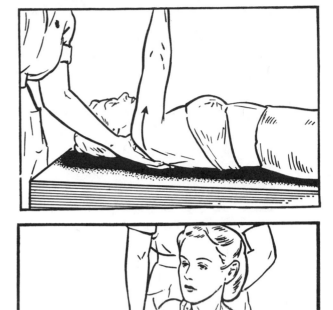

FAIR

Backlying with arm abducted to 90 degrees.
Patient adducts arm to vertical position.

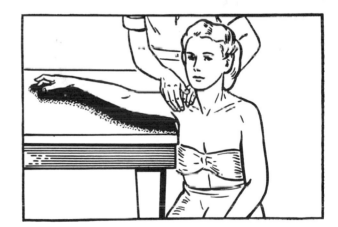

POOR

Sitting with arm resting on table in 90 degrees of abduction.
Stabilize trunk.
Patient brings arm forward through range of motion.

TRACE AND ZERO

Examiner palpates tendon of Pectoralis major near insertion on anterior aspect of upper arm. Muscle fibers of both sternal and clavicular portions may be observed and palpated on upper anterior aspect of thorax. (Latter not illustrated.)

Note: Sternal and clavicular portions of Pectoralis major may be isolated to some degree. In *Normal and Good* tests resistance is given in a direction opposite to line of pull of fibers—i.e., upward and outward for sternal part and downward and outward for clavicular part. In *Fair* test the arm may be placed above 90 degrees of abduction in order to test sternal portion and below 90 degrees to test clavicular portion; arm is then raised to vertical in the direction of muscle fibers being tested. (Not illustrated.)

SHOULDER LATERAL ROTATION

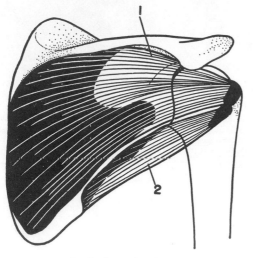

1. Infraspinatus
2. Teres minor

Range of Motion:

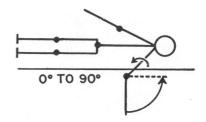

0° TO 90°

Factors Limiting Motion:

1. Tension of superior portion of capsular ligament and coracohumeral ligament
2. Tension of medial rotator muscles of shoulder

Fixation:

Contraction of Trapezius and Rhomboideus major and minor muscles to fix scapula

---

PRIME MOVERS

MUSCLE	ORIGIN	INSERTION
Infraspinatus N: Suprascapular (C5, 6)	a. Medial two thirds of infraspinatus fossa	a. Middle impression on greater tubercle of humerus
Teres minor N: Axillary (C5)	a. Cranial two thirds of axillary border of scapula on dorsal surface	a. Most inferior of 3 impressions on greater tubercle of humerus and area just distal to it, uniting with posterior portion of capsule of shoulder joint

Accessory Muscle

Deltoideus (posterior fibers)

SHOULDER LATERAL ROTATION

NORMAL AND GOOD

Facelying with shoulder abducted to 90 degrees, upper arm supported on table and lower arm hanging vertically over edge.

Stabilize scapula with hand and forearm, but allow freedom for rotation.

Patient swings lower arm forward and upward and laterally rotates shoulder through range of motion. Resistance is given above wrist on forearm.

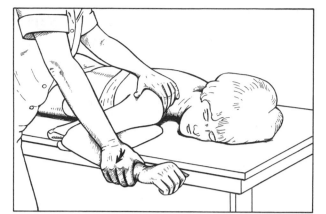

FAIR

Facelying with shoulder abducted to 90 degrees, upper arm supported on table and lower arm hanging vertically over edge.

Stabilize scapula and place hand against anterior surface of arm to prevent abduction (without interfering with motion).

Patient swings lower arm forward and upward and laterally rotates shoulder through range of motion.

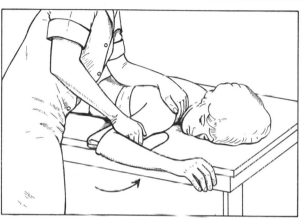

POOR

Facelying with entire arm over edge of table in medially rotated position.

Stabilize scapula.

Patient laterally rotates arm through range of motion.

(Supination of the forearm should not be allowed to substitute for full range in lateral rotation.)

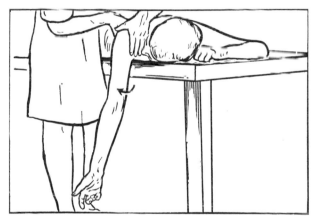

TRACE AND ZERO

The Teres minor may be palpated on axillary border of scapula, and Infraspinatus over body of scapula below the spine.

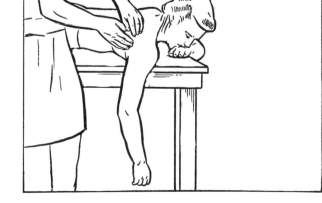

SHOULDER MEDIAL ROTATION

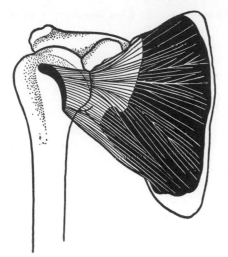

Subscapularis

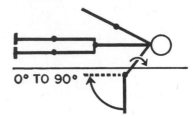

0° TO 90°

Factors Limiting Motion:

   1. Tension of superior portion of capsular ligament

   2. Tension of lateral rotator muscles of shoulder

Fixation:

   1. Weight of trunk

   2. Contraction of Trapezius and Rhomboideus major and minor muscles to fix scapula

PRIME MOVERS

MUSCLE	ORIGIN	INSERTION
Subscapularis N: Upper and lower subscapular (C5, 6)	a. Medial two thirds of costal surface of scapula b. Inferior two thirds of groove on axillary border of scapula	a. Lesser tubercle of humerus b. Ventral portion of capsule of shoulder joint
Pectoralis major (Illus. on p. 98) N: Medial and lateral pectoral (C5, 6, 7, 8, Th1)	a. Anterior surface of sternal half of clavicle b. Half of breadth of ventral surface of sternum as far caudalward as seventh rib c. Cartilages of first 6 or 7 ribs	a. Crest of greater tubercle of humerus
Latissimus dorsi (Illus. on p. 92) N: Thoracodorsal (C6, 7, 8)	a. Spinous processes of lower 6 thoracic vertebrae b. Posterior layer of lumbodorsal fascia by which it is attached to spines of lumbar and sacral vertebrae, supraspinal ligament, and to posterior iliac crest c. External lip of iliac crest lateral to Sacrospinalis d. Three or 4 lower ribs e. Usually a few fibers from inferior angle of scapula	a. Bottom of intertubercular groove of humerus
Teres major (Illus. on p. 92) N: Lowest subscapular (C5, 6)	a. Dorsal surface of inferior angle of scapula	a. Crest below lesser tuberosity of humerus (posterior to Latissimus dorsi)

Accessory Muscle

Deltoideus (anterior fibers)

SHOULDER MEDIAL ROTATION

NORMAL AND GOOD

Facelying with shoulder abducted to 90 degrees, upper arm supported on table and lower arm hanging vertically over edge.

Stabilize scapula with hand and forearm, but allow freedom for rotation.

Patient swings lower arm backward and upward and medially rotates shoulder through range of motion. Resistance is given proximal to wrist on forearm.

FAIR

Facelying with shoulder abducted to 90 degrees, upper arm supported on table and lower arm hanging vertically over edge.

Stabilize scapula.

Patient swings lower arm backward and upward and medially rotates shoulder through range of motion.

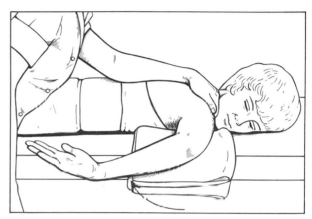

POOR

Facelying with arm over edge of table in lateral rotation.

Stabilize scapula.

Patient medially rotates arm through range of motion.

(Pronation of the forearm should not be allowed to substitute for full range in medial rotation.)

TRACE AND ZERO

Fibers of Subscapularis may be palpated deep in axilla near insertion. (Latissimus dorsi, page 92; Pectoralis major, page 98.)

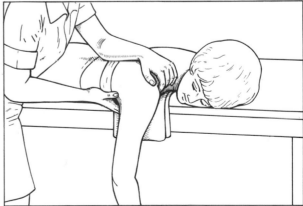

ELBOW FLEXION

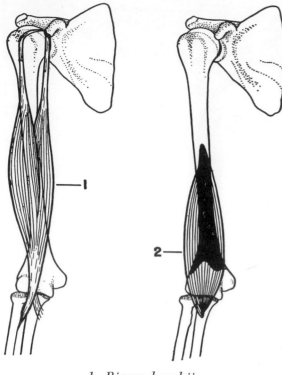

Range of Motion:

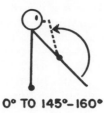

0° TO 145°- 160°

Factors Limiting Motion:

    1. Contact of muscle masses on volar aspect of arm and forearm

    2. Contact of coronoid process with coronoid fossa of humerus

Fixation:

    1. Weight of arm

    2. Fixator muscles of scapula

1. Biceps brachii
2. Brachialis

PRIME MOVERS

MUSCLE	ORIGIN	INSERTION
Biceps brachii N: Musculocutaneous (C5, 6)	*Short head:* a. Flattened tendon from apex of coracoid process of scapula *Long head:* a. Tendon from supraglenoid tuberosity of scapula	a. Passes through capsule of shoulder joint and down in intertubercular groove to insert on posterior portion of tuberosity of radius (Lacertus fibrosus into deep fascia of forearm)

Brachialis
N: Musculocutaneous
   (C5, 6) and usually
   filament from radial
	a. Distal half of anterior aspect of humerus	a. Tuberosity of ulna and anterior surface of coronoid process

Brachioradialis N: Radial (C5, 6)	a. Proximal two-thirds of lateral supracondylar ridge of humerus b. Lateral intermuscular septum distal to groove for radial nerve	a. Flat tendon into lateral side of base of styloid process of radius

Accessory Muscles

Flexor muscles of forearm (arising from medial epicondyle of humerus)

ELBOW FLEXION

NORMAL AND GOOD

Sitting with arm at side and forearm
   supinated.
Stabilize upper arm.
Patient flexes elbow through range of
   motion. Resistance is given proximal to
   wrist joint.

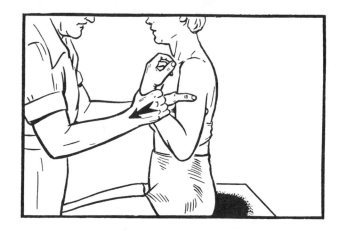

FAIR

Sitting with arm at side and forearm supinated.
Stabilize upper arm.
Patient flexes elbow through range of
   motion.

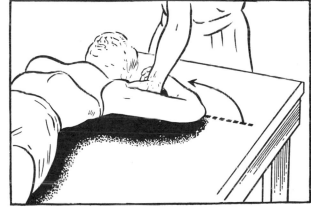

POOR

Backlying with shoulder abducted to 90 degrees and laterally rotated.
Stabilize upper arm.
Patient slides forearm along table through
   complete range of elbow flexion.
(If range of motion is limited in lateral rotation at shoulder joint, test may be given
   with arm medially rotated.)

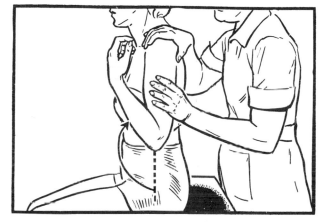

TRACE AND ZERO

Examiner may palpate tendon of Biceps
   brachii in antecubital space; muscle fibers
   may be found on anterior surface of arm.
   (Latter not illustrated.)

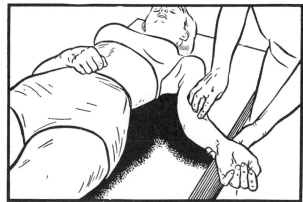

ELBOW FLEXION

Note: If patient attempts to substitute with Brachioradialis, forearm will be held in midposition between pronation and supination. If Pronator teres substitutes, the forearm will be pronated. (Latter not illustrated.)

Note: The wrist flexors may be contracted for assistance in elbow flexion. Wrist will be strongly flexed as a result. Wrist should be relaxed.

FOR NOTES:

ELBOW EXTENSION

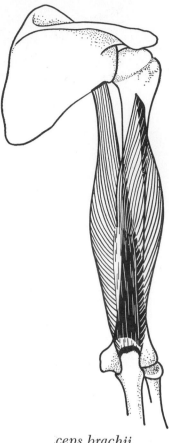

ceps brachii

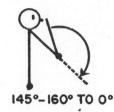

145°– 160° TO 0°

Factors Limiting Motion:

1. Tension of anterior, radial and ulnar collateral ligaments of elbow joint
2. Tension of flexor muscles of forearm
3. Contact of olecranon process with olecranon fossa on posterior aspect of humerus

Fixation:

1. Weight of arm
2. Contraction of fixator muscles of scapula

PRIME MOVERS

MUSCLE	ORIGIN	INSERTION
Triceps brachii N: Radial (C7, 8)	*Long head:* a. Infraglenoid tuberosity of scapula *Lateral head:* a. Posterior surface of body of humerus proximal to groove for radial nerve *Medial head (Deep head):* a. Posterior surface of body of humerus distal to groove for radial nerve	a. Posterior portion of proximal surface of olecranon b. Fibrous expansion to deep fascia of forearm

Accessory Muscles

Anconeus
Extensor muscles of forearm (arising from lateral condyle of humerus)

ELBOW EXTENSION

NORMAL AND GOOD

Backlying with shoulder flexed to 90 degrees and elbow flexed.
Stabilize arm.
Patient extends elbow through range of motion. Resistance is given proximal to wrist joint in plane of forearm motion.

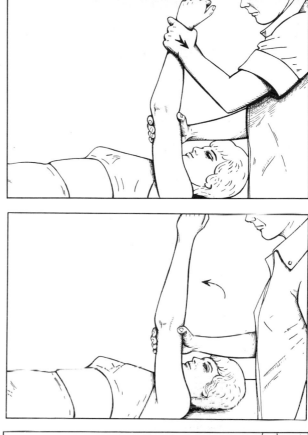

FAIR

Backlying with shoulder flexed to 90 degrees and elbow flexed.
Stabilize arm.
Patient extends elbow through range of motion.

POOR

Backlying with arm abducted to 90 degrees and laterally rotated. Elbow is flexed.
Stabilize arm.
Patient extends elbow through range of motion. (If range of motion is limited in lateral rotation at shoulder joint, test may be given with arm medially rotated.)

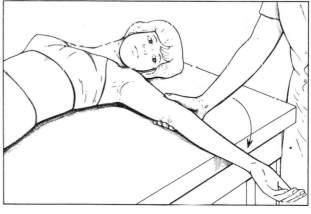

TRACE AND ZERO

Examiner may palpate tendon of Triceps brachii at the elbow joint and muscle fibers on posterior surface of arm. (Latter not illustrated.)

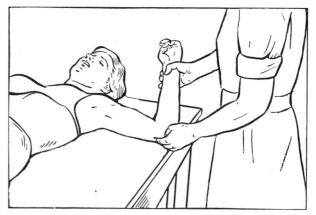

FOREARM SUPINATION

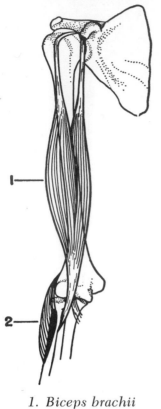

1. *Biceps brachii*
2. *Supinator*

Range of Motion:

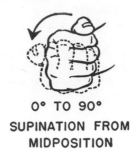

0° TO 90°

SUPINATION FROM MIDPOSITION

Factors Limiting Motion:

1. Tension of volar radioulnar ligament and ulnar collateral ligament of wrist joint
2. Tension of oblique cord and lowest fibers of interosseous membrane
3. Tension of pronator muscles of forearm.

Fixation:

Weight of arm

PRIME MOVERS

MUSCLE	ORIGIN	INSERTION
Biceps brachii N: Musculocutaneous (C5, 6)	*Short head:* a. Flattened tendon from apex of coracoid process of scapula *Long head:* a. Tendon from supraglenoid tuberosity	a. Posterior portion of tuberosity of radius (Lacertus fibrosus into deep fascia of forearm)
Supinator N: Radial (C6)	a. Lateral epicondyle of humerus b. Ridge and depression on ulna distal to radial notch c. Annular ligament and radial collateral ligament of elbow joint	a. Muscle winds around radius to insert into dorsal and lateral surfaces of body of radius between oblique line and head of bone

Accessory Muscle

Brachioradialis

FOREARM SUPINATION

NORMAL AND GOOD

Sitting with arm at side, elbow flexed to 90 degrees and forearm pronated to prevent rotation at the shoulder. Muscles of wrist and fingers are relaxed.

Stabilize arm.

Patient supinates forearm. Resistance is given on dorsal surface of distal end of radius with counterpressure against the ventral surface of the ulna.

(Resistance may be given by grasping around the dorsal surface of the hand instead of the position illustrated.)

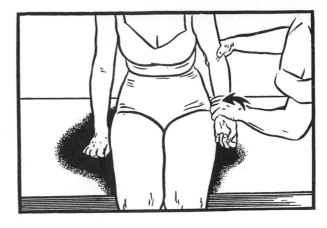

FAIR AND POOR

Sitting with arm at side, elbow flexed to 90 degrees, forearm pronated and supported by examiner. Muscles of wrist and fingers are relaxed.

Patient supinates forearm through full range of motion for fair grade and through partial range for poor grade.

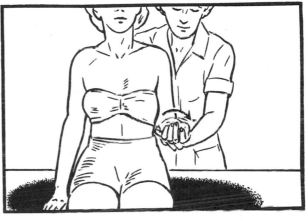

TRACE AND ZERO

Supinator muscle is palpable on radial side of forearm if overlying extensor muscles are not functioning. Tendon of Biceps brachii is found in antecubital space.

Note: Patient should not be allowed to laterally rotate arm and move elbow across thorax as forearm is supinated. As a result of this movement the forearm may appear to be supinated, but range of motion is incomplete. This motion may "roll" the forearm into supination without a muscular contraction taking place.

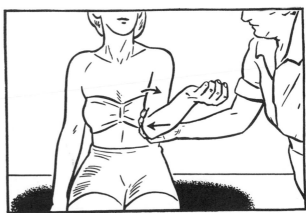

FOREARM PRONATION

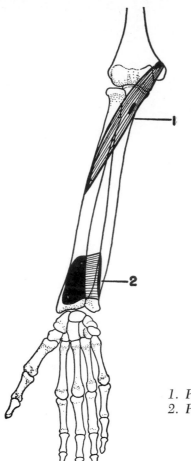

Range of Motion:

0° TO 90°
PRONATION FROM
MIDPOSITION

Factors Limiting Motion:

    1. Tension of dorsal radioulnar, ulnar collateral and dorsal radiocarpal ligaments
    2. Tension of lowest fibers of interosseous membrane

Fixation:

    Weight of arm

1. Pronator teres
2. Pronator quadratus

PRIME MOVERS

MUSCLE	ORIGIN	INSERTION
Pronator teres N: Median (C6, 7)	*Humeral head·* a. Area proximal to medial epicondyle of humerus b. Common tendon of flexor muscle group *Ulnar head:* a. Medial side of coronoid process of ulna (joins humeral head at acute angle)	a. Rough impression on lateral surface of radius at middle of body
Pronator quadratus N: Median (Palmar interosseous branch) (C8, Th1)	a. Palmar surface of lower fourth of ulna	a. Distal fourth of lateral border and palmar surface of radius b. Deeper fibers into triangular area proximal to ulnar notch of radius

Accessory Muscle

Flexor carpi radialis

FOREARM PRONATION

NORMAL AND GOOD

Sitting with arm at side, elbow flexed to 90 degrees to prevent rotation at the shoulder and forearm supinated. Muscles of wrist and fingers are relaxed.

Stabilize arm.

Patient pronates forearm through range of motion. Resistance is given on volar surface of distal end of radius with counter-pressure against the dorsal surface of the ulna for derotation.

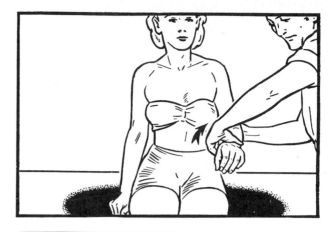

FAIR AND POOR

Sitting with arm at side, elbow flexed to 90 degrees, forearm supinated and supported by examiner. Muscles of wrist and fingers are relaxed.

Patient pronates forearm through full range of motion for fair grade and through partial range for poor grade.

TRACE AND ZERO

Sitting.

Examiner palpates fibers of Pronator teres on upper third of volar surface of forearm on a diagonal line from medial condyle of humerus to lateral border of radius.

Note: Patient should not be allowed to medially rotate or abduct upper arm during pronation. This movement makes the range of motion in pronation appear complete and allows forearm to roll into pronated position.

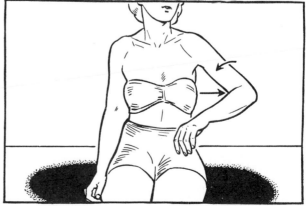

WRIST FLEXION

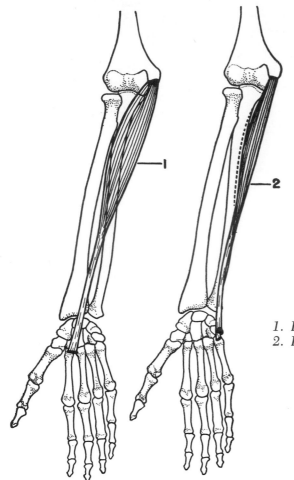

Range of Motion:

Wrist flexion: 0 to 90 degrees

Factors Limiting Motion:

Tension of dorsal radiocarpal ligament

Fixation:

Weight of arm

1. *Flexor carpi radialis*
2. *Flexor carpi ulnaris*

PRIME MOVERS

Muscle	Origin	Insertion
Flexor carpi radialis N: Medial (C6, 7)	a. Medial epicondyle of humerus by common tendon	a. Base of second metacarpal bone on palmar surface b. May send slip to base of third metacarpal bone
Flexor carpi ulnaris N: Ulnar (C8, Th1)	a. Medial epicondyle of humerus by common tendon (humeral head) b. Medial margin of olecranon process and upper two thirds of dorsal border of ulna (ulnar head)	a. Pisiform bone b. Prolongations to hamate and base of fifth metacarpal bone

Accessory Muscle

Palmaris longus

WRIST FLEXION

NORMAL AND GOOD

Sitting with forearm resting on table with
forearm supinated. Muscles of thumb and
fingers relaxed.

Stabilize forearm.

Patient flexes wrist.

To test Flexor carpi radialis, resistance is
given at base of second metacarpal bone
in direction of extension and ulnar deviation (Illustrated).

To test Flexor carpi ulnaris, resistance is
given at base of fifth metacarpal bone in
direction of extension and radial deviation. (Not illustrated.)

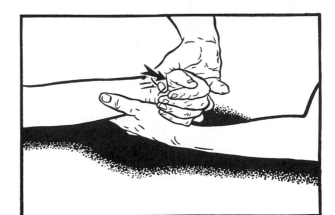

FAIR

Sitting with forearm resting on table with
forearm supinated. Muscles of thumb and
fingers relaxed.

Stabilize forearm.

Patient flexes wrist with radial deviation or
ulnar deviation.

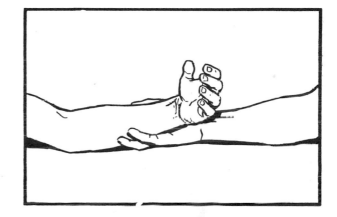

POOR

Sitting, forearm supported, hand resting on
medial border. Muscles of thumb and
fingers relaxed.

Stabilize forearm.

Patient flexes wrist, sliding hand along
table. Deviation should be observed and
muscles graded accordingly.

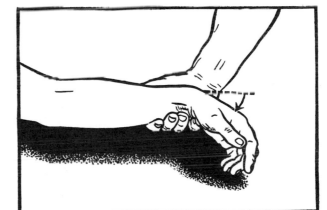

TRACE AND ZERO

Examiner palpates tendon of Flexor carpi
radialis on lateral palmar aspect of wrist
and tendon of Flexor carpi ulnaris on
medial palmar surface.

★ ★ ★ ★ ★

Note: Do not allow "curling" of fingers
prior to flexion or during resistance to
prevent substitution by finger flexors.

WRIST EXTENSION

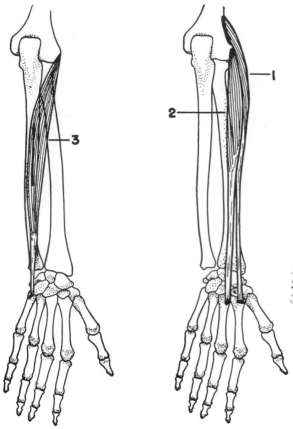

Range of Motion:

Wrist extension beyond midline: 0 to 70 degrees

Factors Limiting Motion:

Tension of palmar radiocarpal ligament

Fixation:

Weight of arm

1. *Extensor carpi radialis longus*
2. *Extensor carpi radialis brevis*
3. *Extensor carpi ulnaris*

PRIME MOVERS

MUSCLE	ORIGIN	INSERTION
Extensor carpi radialis longus N: Radial (C6, 7)	a. Distal third of lateral supracondylar ridge of humerus b. Some fibers from common extensor tendon of lateral epicondyle of humerus	a. Dorsal surface of base of second metacarpal bone on radial side
Extensor carpi radialis brevis N: Radial (C6, 7)	a. Lateral epicondyle of humerus by common tendon b. Radial collateral ligament of elbow joint	a. Dorsal surface of base of third metacarpal bone on radial side
Extensor carpi ulnaris N: Radial (C6, 7, 8)	a. Lateral epicondyle of humerus by common extensor tendon b. Aponeurosis from dorsal border of ulna	a. Tubercle on ulnar side of base of fifth metacarpal bone

WRIST EXTENSION

NORMAL AND GOOD

Sitting with forearm resting on table with
forearm pronated. Muscles of fingers and
thumb relaxed.

Stabilize forearm.

Patient extends wrist.

To test Extensor carpi radialis longus and
brevis, resistance is given on dorsal surface of second and third metacarpal bones
in direction of flexion and ulnar deviation.

To test Extensor carpi ulnaris, resistance is
given on dorsal surface of fifth metacarpal bone in direction of flexion and
radial deviation.

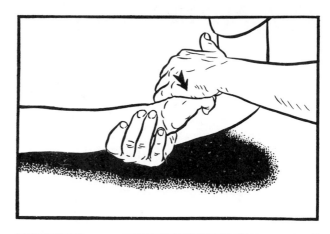

FAIR

Sitting with forearm resting on table with
forearm pronated. Muscles of fingers and
thumb relaxed.

Stabilize forearm.

Patient extends wrist with radial deviation
or ulnar deviation.

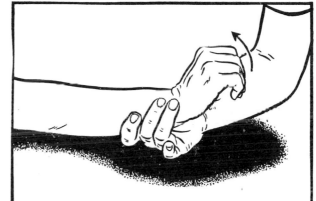

POOR

Sitting, forearm supported, hand resting on
medial border.

Stabilize forearm.

Patient extends wrist, sliding hand along
table through range of motion.

Deviation should be observed and muscles
graded accordingly.

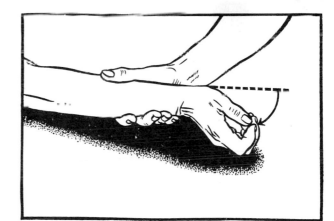

TRACE AND ZERO

Tendons of wrist extensors may be found on
lateral dorsal surface of wrist in line with
second and third metacarpal bones and
on medial dorsal surface proximal to fifth
metacarpal bone.

★ ★ ★ ★ ★

Note: Do not allow extension of fingers
prior to extension or during resistance to
prevent substitution by finger extensors.

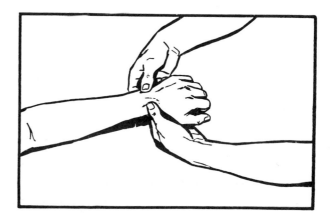

FLEXION OF METACARPOPHALANGEAL JOINTS OF FINGERS

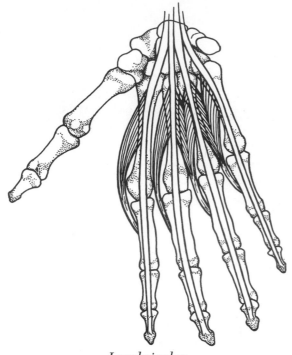

Lumbricales

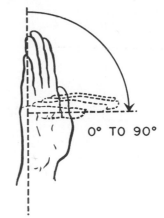

Range of Motion:

0° TO 90°

Factors Limiting Motion:

Tension of expansions of extensor tendons of fingers.

Fixation:

Weight of arm

PRIME MOVERS

MUSCLE	ORIGIN	INSERTION
Lumbricales Third and fourth lumbricales N: Ulnar (C8) First and second lumbricales N: Median (C6, 7)	a. Arise from tendons of Flexor digitorum profundus muscle *First and second:* Radial sides and palmar surfaces of tendons of index and middle fingers *Third:* Adjacent sides of tendons of middle and ring fingers *Fourth:* Adjacent sides of tendons of ring and little fingers	a. Pass to radial side of corresponding fingers to insert opposite metacarpophalangeal joints into tendinous expansions of Extensor digitorum communis, covering dorsal aspect of fingers
Interossei dorsales (Illustrated on p. 124) N: Ulnar (C8)	a. Each by 2 heads from adjacent sides of metacarpal bones between which it lies	a. Base of proximal phalanges of 4 fingers: first and second into radial side of index and middle fingers; third and fourth into ulnar side of middle and ring fingers.
Interossei palmares (Illustrated on p. 126) N: Ulnar (C8, T1)	a. Entire length of second, fourth and fifth metacarpal bones on palmar surface	a. Side of base of proximal phalanx of corresponding finger; first into ulnar side of index finger; second and third into radial side of ring and little fingers b. Into aponeurotic expansion of Extensor digitorum communis tendons of same fingers

(Continued on page 119.)

FLEXION OF METACARPOPHALANGEAL JOINTS OF FINGERS

NORMAL AND GOOD

Sitting with hand resting on dorsal surface. Stabilize metacarpals.

Patient flexes fingers at metacarpophalangeal joints, keeping interphalangeal joints extended. Resistance is given on palmar surface of proximal row of phalanges.

Note: Resistance may be given to each finger separately if Lumbricales are unequal in strength.

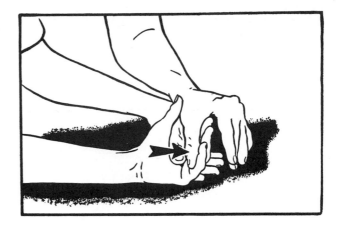

FAIR AND POOR

Sitting with hand supported. Stabilize metacarpals.

Patient flexes fingers at metacarpophalangeal joints through range of motion, keeping interphalangeal joints extended.

Patient flexes metacarpophalangeal joints through full range of motion for fair grade and through partial range for poor grade.

(Forearm partially pronated in illustration to show range of motion.)

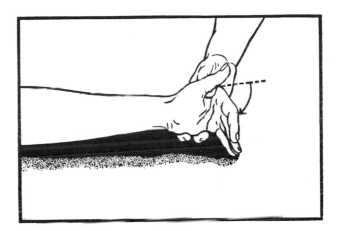

TRACE AND ZERO

Contraction of Lumbricales may be detected by light pressure against palmar surface of proximal phalanges as patient attempts to flex at metacarpophalangeal joints.

★ ★ ★ ★ ★

Note: The Flexor digitorum superficialis and Flexor digitorum profundus should not be allowed to substitute for Lumbricales with flexion of fingers. These muscles should be kept relaxed as much as possible with motion limited to metacarpophalangeal joint. (Not illustrated.)

Note: Individual testing of fingers (in all tests) is often desirable as they vary in strength.

MUSCLE	ORIGIN	INSERTION
Accessory Muscles		
Flexor digiti minimi brevis		
Flexor digitorum superficialis		
Flexor digitorum profundus		

FLEXION OF PROXIMAL AND DISTAL INTERPHALANGEAL JOINTS OF FINGERS

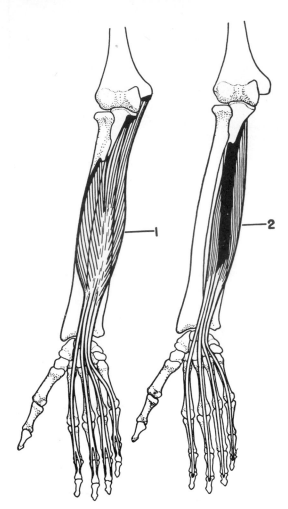

1. *Flexor digitorum superficialis*
2. *Flexor digitorum profundus*

Range of Motion:

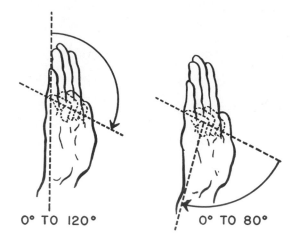

0° TO 120°   0° TO 80°

Factors Limiting Motion:

Tension of expansions of Extensor digitorum communis tendons (dorsal ligaments)

Fixation:

1. Weight of arm
2. Synergistic action of wrist extensor muscles to prevent flexion of wrist

Flexion of Proximal Interphalangeal Joints of Fingers

PRIME MOVERS

MUSCLE	ORIGIN	INSERTION
Flexor digitorum superficialis N: Median (C7, 8, Th1)	a. Medial epicondyle of humerus by common flexor tendon (humeral head) b. Medial side of coronoid process of ulna (ulnar head) c. Oblique line of radius from radial tuberosity to insertion of Pronator teres (radial head)	*Tendon divides:* a. Superficial part into middle and ring fingers (insertions into sides of second phalanges) b. Deep part into index and little fingers (insertions into sides of second phalanges)

FLEXION OF PROXIMAL AND DISTAL INTERPHALANGEAL JOINTS OF FINGERS

FLEXION OF PROXIMAL INTERPHALANGEAL JOINTS OF FINGERS (FLEXOR DIGITORUM SUPERFICIALIS)

NORMAL AND GOOD

Sitting with hand resting on dorsal surface, palm upward with wrist and fingers extended.

Stabilize proximal phalanx of finger.

Patient flexes middle phalanx, and resistance is given on palmar surface.

FAIR AND POOR

Patient flexes proximal phalanx through full range of motion for fair grade and through partial range for poor grade. (Not illustrated.)

FLEXION OF DISTAL INTERPHALANGEAL JOINTS OF FINGERS (FLEXOR DIGITORUM PROFUNDUS)

NORMAL AND GOOD

Sitting with hand resting palm upward on table and fingers extended.

Stabilize middle phalanx of finger.

Patient flexes distal phalanx. Resistance is given on palmar surface of distal phalanx of finger.

FAIR AND POOR

Patient flexes distal phalanx through full range of motion for fair grade and through partial range for poor grade. (Not illustrated.)

TRACE AND ZERO

Flexor digitorum profundus may be palpated on the palmar surface of the middle phalanx. (Not illustrated.)

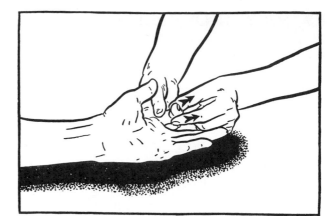

TRACE AND ZERO

Superficial portion of the Flexor digitorum superficialis may be palpated at the wrist under the Palmaris longus. (Not illustrated.)

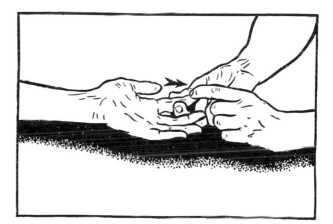

Note: In testing flexion of proximal interphalangeal joints, if Flexor digitorum profundus is allowed to contract, all joints of fingers will be flexed. Relaxation may be obtained by showing patient exactly where action takes place as a result of contraction of Flexor digitorum superficialis and moving joint through range of motion passively several times before an active contraction is allowed. (Not illustrated.)

Flexion of Distal Interphalangeal Joints of Fingers

PRIME MOVERS

MUSCLE	ORIGIN	INSERTION
Flexor digitorum profundus N: Ulnar (C8, Th1) and median (palmar interosseus branch)	a. Proximal three fourths of volar and medial surfaces of ulna b. Aponeurosis from upper three fourths of dorsal border of ulna c. Medial side of coronoid process	a. Bases of distal phalanges of 4 fingers (tendons pass through those of Flexor digitorum superficialis)

EXTENSION OF METACARPOPHALANGEAL JOINTS OF FINGERS

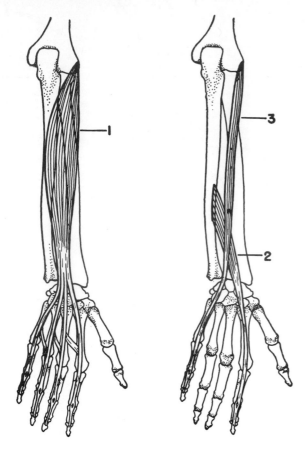

1. *Extensor digitorum communis*
2. *Extensor indicis proprius*
3. *Extensor digiti minimi*

Range of Motion:

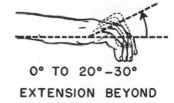

0° TO 20°–30°
EXTENSION BEYOND
MIDLINE

Factors Limiting Motion:

1. Tension of palmar and collateral ligaments
2. Tension of flexor muscles of fingers

Fixation:

1. Weight of arm
2. Synergistic action of wrist flexor muscles to prevent wrist extension

PRIME MOVERS

MUSCLE	ORIGIN	INSERTION
Extensor digitorum communis N: Radial (C6, 7, 8)	a. Lateral epicondyle of humerus by common tendon	a. Four tendons into base of second and third phalanges of fingers (Opposite metacarpophalangeal joints, tendons are bound by fasciculi to collateral ligaments)
Extensor indicis proprius N: Radial (C6, 7, 8)	a. Dorsal surface of body of ulna below origin of Extensor pollicis longus	a. Joins ulnar side of tendon of Extensor digitorum communis, which goes to index finger; terminates in extensor expansion

(Continued on page 123.)

EXTENSION OF METACARPOPHALANGEAL JOINTS OF FINGERS

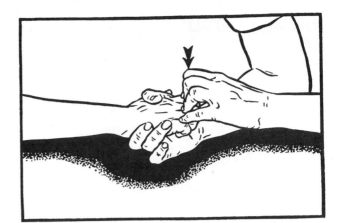

NORMAL AND GOOD

Arm resting on table, hand supported, wrist in midposition, fingers flexed.

Stabilize metacarpals.

Patient extends proximal row of phalanges with interphalangeal joints partially flexed. Resistance is given on dorsal surface of proximal row of phalanges of fingers.

Note: Resistance may be given to each finger separately. Extensor indicis proprius assists in extension of index finger, and Extensor digiti quinti proprius assists in extension of fifth finger.

FAIR AND POOR

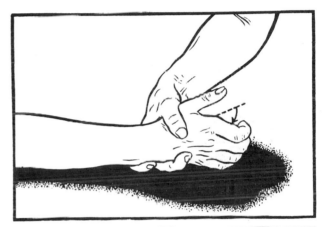

Sitting with hand supported, fingers flexed and wrist in midposition.

Stabilize metacarpals.

Patient extends proximal row of phalanges to end of range, with interphalangeal joints partially flexed.

Patient extends metacarpophalangeal joints through full range of motion for grade of fair and through partial range for grade of poor.

TRACE AND ZERO

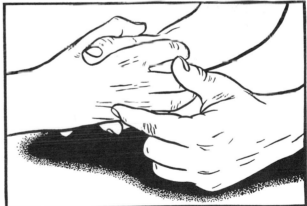

The tendons of the finger extensors may easily be located on dorsum of hand where they pass over metacarpals.

PRIME MOVERS

MUSCLE	ORIGIN	INSERTION
Extensor digiti minimi N: Radial (C7)	a. Common extensor tendon from lateral epicondyle of humerus	a. Joins expansion of Extensor digitorum communis tendon on dorsum of the first phalanx of fifth finger

FINGER ABDUCTION

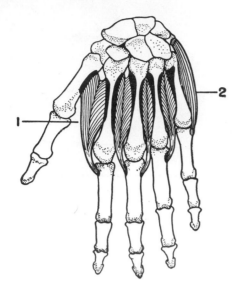

1. *Interossei dorsales*
2. *Abductor digiti minimi*

Range of Motion:

0° TO 20°–25°

Factors Limiting Motion:

Tension of fascia and skin between fingers

Fixation:

1. Contraction of Flexor carpi ulnaris muscle
2. Weight of hand

PRIME MOVERS

MUSCLE	ORIGIN	INSERTION
Interossei dorsales N: Ulnar (C8, T1)	a. Each by 2 heads from adjacent sides of metacarpal bones between which it lies	a. Base of proximal phalanges of 3 fingers: first and second into radial side of index and middle fingers; third and fourth into ulnar side of middle and ring fingers b. Into aponeurotic expansions of Extensor digitorum tendons of corresponding fingers
Abductor digiti minimi N: Ulnar (C8)	a. Pisiform bone b. Tendon of Flexor carpi ulnaris	*Tendon divides:* a. One part into ulnar side of first phalanx of little finger at base b. Other part into ulnar border of aponeurosis of Extensor digiti minimi muscle

FINGER ABDUCTION

NORMAL AND GOOD

(Test for first and third Interossei dorsales)

Sitting with hand supported palm downward, fingers adducted.

Stabilize metacarpals.

Patient abducts fingers. Resistance is given on radial side of second and ulnar side of third finger. (To test individual fingers, resistance is given on first phalanx.)

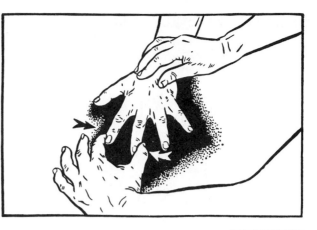

NORMAL AND GOOD

(Test for second and fourth Interossei dorsales and Abductor digiti minimi)

Patient abducts fingers. Resistance is given on ulnar side of fourth and fifth fingers and on radial side of third finger.

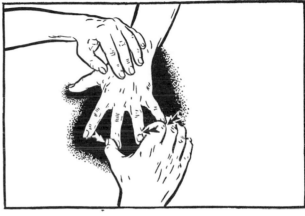

FAIR AND POOR

Sitting with palm resting on table, fingers adducted.

Patient abducts fingers through range of motion. (Third finger must be moved in both directions.)

Patient abducts fingers through full range of motion for fair grade and through partial range for poor grade.

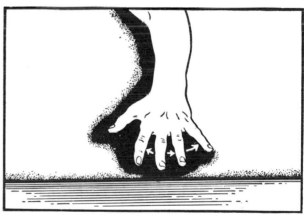

TRACE AND ZERO

The Interossei dorsales lie deep between the metacarpal bones on the dorsum of the hand. (Palpation of first Interosseus dorsales shown in illustration.)

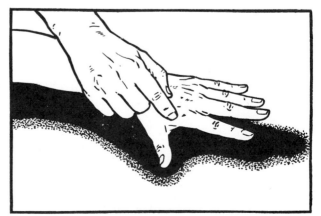

FINGER ADDUCTION

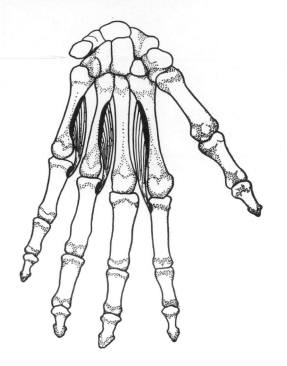

Interossei palmares

Range of Motion:

20°- 25° TO 0°

Factors Limiting Motion:

Contact of fingers

Fixation:

Weight of hand

PRIME MOVER

MUSCLE	ORIGIN	INSERTION
Interossei palmares N: Ulnar (C8, T1)	a. Entire length of second, fourth and fifth metacarpal bones on palmar surface	a. Side of base of proximal phalanx of corresponding finger: first into ulnar side of index finger; second and third into radial side of ring and little fingers b. Into aponeurotic expansion of Extensor digitorum tendon of same finger

FINGER ADDUCTION

NORMAL AND GOOD

Sitting with hand supported palm downward, fingers abducted.

Patient adducts fingers. Resistance is given in radial direction on second finger and in ulnar direction on fourth and fifth fingers.

FAIR AND POOR

Sitting with hand resting palm downward on table, fingers in abduction.

Patient adducts fingers through full range of motion for fair grade and through partial range for poor grade.

TRACE AND ZERO

Presence of contraction of the Interossei palmares may be determined by outward pressure on the second, fourth and fifth fingers as the patient attempts to adduct.

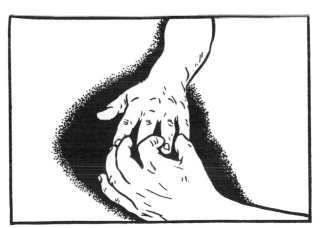

FLEXION OF METACARPOPHALANGEAL AND INTERPHALANGEAL JOINTS OF THUMB

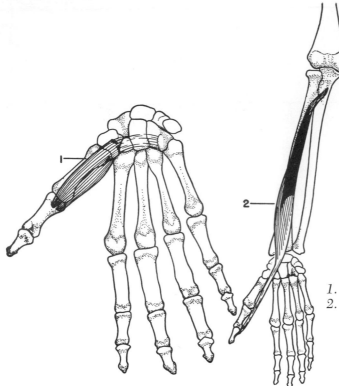

Range of Motion:

Metacarpophalangeal flexion: 0 to 60–70 degrees

Interphalangeal flexion: 0 to 90 degrees

Factors Limiting Motion:

Tension of tendons of extensor muscles of thumb

Fixation:

Weight of forearm and hand

1. *Flexor pollicis brevis*
2. *Flexor pollicis longus*

Flexion of Metacarpophalangeal Joint of Thumb
PRIME MOVER

MUSCLE	ORIGIN	INSERTION
Flexor pollicis brevis (lateral portion) N: Median (C6, 7)	*Lateral portion* (superficial): a. Distal border of flexor retinaculum b. Ridge on trapezium bone	a. Base of proximal phalanx of thumb on radial side (sesamoid bone)
(medial portion) N: Ulnar (C8, T1)	*Medial portion* (deep): a. Ulnar side of first metacarpal bone between Adductor pollicis (obliquus) and lateral head of first Interosseus dorsales	a. Ulnar side of base of first phalanx of thumb with Adductor pollicis (obliquus)

Flexion of Interphalangeal Joint of Thumb
PRIME MOVER

MUSCLE	ORIGIN	INSERTION
Flexor pollicis longus N: Median (C8, Th1)	a. Volar surface of body of radius from tuberosity to attachment of Pronator quadratus b. Interosseous membrane c. Usually from coronoid process or medial epicondyle of humerus	a. Base of distal phalanx of thumb

FLEXION OF METACARPOPHALANGEAL AND INTERPHALANGEAL JOINTS OF THUMB

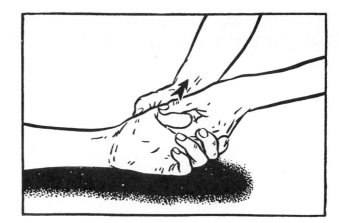

FLEXION OF METACARPOPHALANGEAL JOINT OF THUMB

NORMAL AND GOOD

Sitting with hand resting palm upward on table.

Stabilize first metacarpal.

Patient flexes first phalanx of thumb. Distal phalanx remains relaxed. Resistance is given on palmar surface of proximal phalanx.

FAIR AND POOR

Patient flexes first phalanx of thumb through full range of motion for fair grade and through partial range for poor grade. (Not illustrated.)

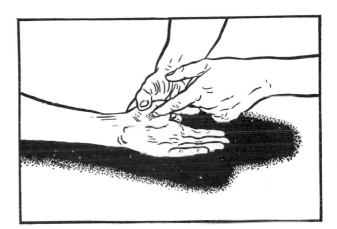

TRACE AND ZERO

Contraction of Flexor pollicis brevis may be determined by pressure over palmar surface of first metacarpal (medial to Abductor pollicis brevis) as patient attempts flexion.

FLEXION OF INTERPHALANGEAL JOINT OF THUMB

NORMAL AND GOOD

Sitting with hand resting palm upward on table.

Stabilize first phalanx of thumb.

Patient flexes distal phalanx (motion takes place in plane of palm). Resistance is given on palmar surface of distal phalanx of thumb.

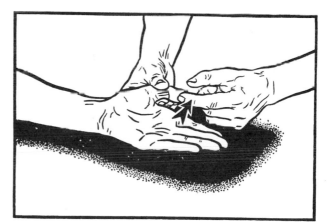

FAIR AND POOR

Patient flexes distal phalanx through full range of motion for fair grade and through partial range for poor grade. (Not illustrated.)

TRACE AND ZERO

The tendon of Flexor pollicis longus may be found on palmar surface of the first phalanx of the thumb. (Illustrated above.)

EXTENSION OF METACARPOPHALANGEAL AND INTERPHALANGEAL JOINTS OF THUMB

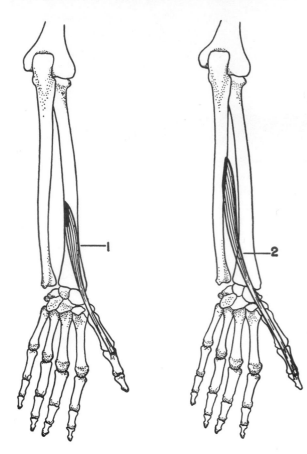

Range of Motion:

Metacarpophalangeal extension: 60–70 to 0 degrees

Interphalangeal extension: 90 to 0 degrees

Factors Limiting Motion:

Tension of palmar and collateral ligaments of thumb

Fixation:

1. Weight of forearm and hand
2. Synergistic action of wrist adductor muscles to prevent wrist abduction (radial flexion)

1. *Extensor pollicis brevis*
2. *Extensor pollicis longus*

Extension of Metacarpophalangeal Joint of Thumb

PRIME MOVER

Muscle	Origin	Insertion
Extensor pollicis brevis N: Radial (C6, 7)	a. Dorsal surface of body of radius distal to Abductor pollicis longus muscle b. Interosseous membrane	a. Base of first phalanx of thumb on dorsal aspect

Extension of Interphalangeal Joint of Thumb

PRIME MOVER

Muscle	Origin	Insertion
Extensor pollicis longus N: Radial (C6, 7, 8)	a. Lateral part of middle third of body of ulna on dorsal surface below Abductor pollicis longus muscle b. Interosseous membrane	a. Base of distal phalanx of thumb on dorsal aspect

EXTENSION OF METACARPOPHALANGEAL AND INTERPHALANGEAL JOINTS OF THUMB

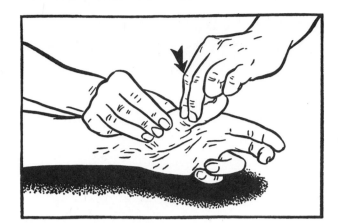

EXTENSION OF METACARPOPHALANGEAL JOINT OF THUMB

NORMAL AND GOOD

Sitting with hand resting on table.
Stabilize first metacarpal.
Patient extends first phalanx of thumb. Resistance is given on dorsal surface of proximal phalanx.

FAIR AND POOR

Patient extends first phalanx of thumb through full range of motion for fair and through partial range for poor. (Not illustrated.)

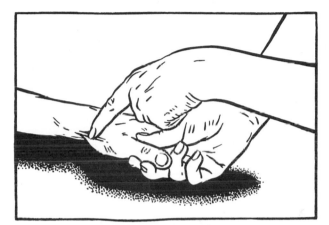

TRACE AND ZERO

Tendon of Extensor pollicis brevis may be found at base of metacarpal of thumb.

EXTENSION OF INTERPHALANGEAL JOINT OF THUMB

NORMAL AND GOOD

Sitting with hand resting on ulnar border.
Stabilize first phalanx of thumb.
Patient extends distal phalanx (motion takes place in plane of palm). Resistance is given on dorsal surface of distal phalanx of thumb.

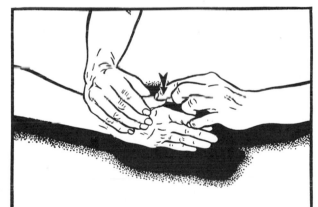

FAIR AND POOR

Patient extends distal phalanx of thumb through full range of motion for fair, and through partial range for poor. (Not illustrated.)

TRACE AND ZERO

Tendon of Extensor pollicis longus may be palpated on dorsal surface of hand between head of first metacarpal and base of second. It may also be found on dorsal surface of first phalanx.

THUMB ABDUCTION

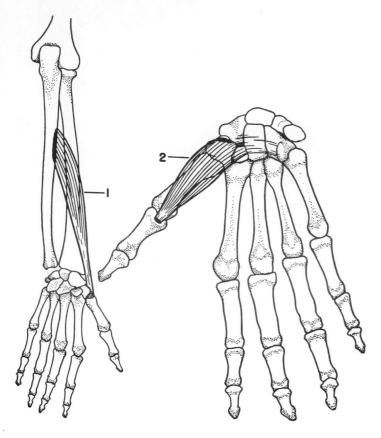

Range of Motion:

0 to 40–50 degrees
(Motion takes place primarily at carpometacarpal joint)

Factors Limiting Motion:

1. Tension of skin between thumb and index finger
2. Tension of first Interosseus dorsales muscle

Fixation:

1. Weight of forearm and hand
2. Synergistic action of wrist adductors to prevent wrist abduction (radial flexion)

1. *Abductor pollicis longus*
2. *Abductor pollicis brevis*

PRIME MOVERS

MUSCLE	ORIGIN	INSERTION
Abductor pollicis longus N: Radial (C6, 7)	a. Lateral part of dorsal surface of body of ulna below Anconeus muscle b. Middle third of dorsal surface of body of radius	a. Radial side of base of first metacarpal bone
Abductor pollicis brevis N: Median (C6, 7)	a. Tuberosity of scaphoid bone b. Ridge of trapezium c. Transverse carpal ligament	a. Radial side of base of first phalanx of thumb b. Capsule of first metacarpophalangeal joint

Accessory Muscle

Palmaris longus

THUMB ABDUCTION

NORMAL AND GOOD

Sitting with hand supported.
Stabilize medial four metacarpals and wrist.
Patient raises thumb vertically through range of abduction. Resistance is given on lateral border of first phalanx of thumb. (If Abductor pollicis longus is stronger than the brevis, thumb will deviate toward radial side of hand. If Abductor pollicis brevis is stronger, deviation will be toward ulnar side.)

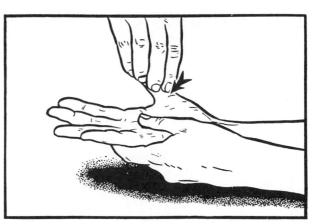

FAIR AND POOR

Sitting with hand supported.
Stabilize metacarpals and wrist.
Patient abducts thumb through full range of motion for fair grade and through partial range for poor grade.

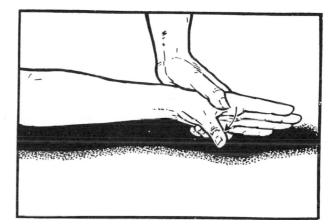

TRACE AND ZERO

The Abductor pollicis brevis fibers may easily be found on thenar eminence lateral to the Flexor pollicis brevis. The tendon of the Abductor pollicis longus may be palpated near its insertion. (Latter not illustrated.)

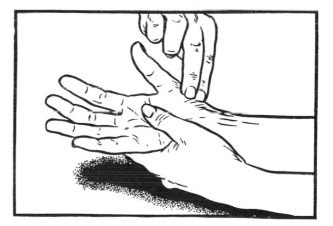

THUMB ADDUCTION

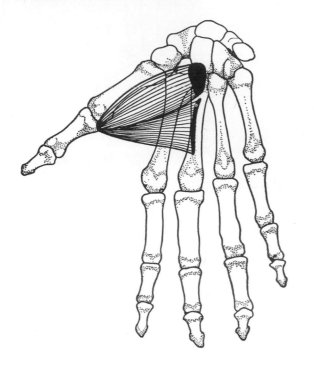

Range of Motion:

40–50 to 0 degrees
(Motion takes place primarily at carpo-metacarpal joint)

Factors Limiting Motion:

Contact of thumb with second metacarpal bone

Fixation:

Weight of hand

Adductor pollicis, caput transversus and caput obliquus

PRIME MOVERS

MUSCLE	ORIGIN	INSERTION
Adductor pollicis, caput obliquus N: Ulnar (C8, T1)	a. Capitate bone b. Base of second and third metacarpal bones on palmar surface c. Intercarpal ligaments	a. Unites with tendon of Flexor pollicis brevis and Adductor pollicis transversus to insert into base of first phalanx of thumb on ulnar side (sesamoid bone) b. Fasciculus to lateral portion of Flexor pollicis brevis and Abductor pollicis brevis
Adductor pollicis, caput transversus N: Ulnar (C8, T1)	a. Distal two thirds of palmar surface of third metacarpal bone	a. Medial part of Flexor pollicis brevis and Adductor pollicis obliquus into base of first phalanx of thumb on ulnar side

THUMB ADDUCTION

NORMAL AND GOOD

Sitting with hand supported.
Stabilize medial four metacarpals.
Patient adducts thumb. Resistance is given
   on medial border of first phalanx.

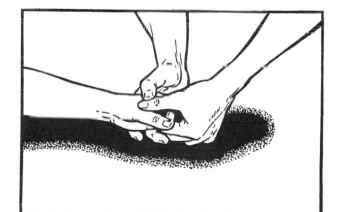

FAIR AND POOR

Sitting with hand supported.
Stabilize metacarpals.
Patient adducts thumb through full range of
   motion for fair grade and through partial
   range for poor grade.

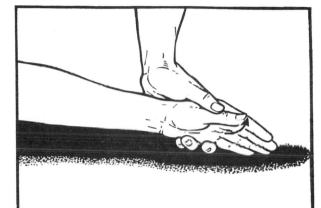

TRACE AND ZERO

Muscle fibers may be palpated between
   first Interossei dorsales muscle and first
   metacarpal bone.

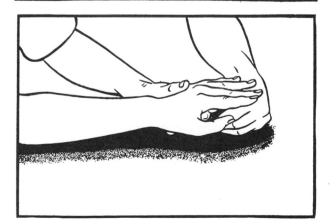

Note: Flexor pollicis longus and Flexor
   pollicis brevis may help pull thumb
   toward palm. These muscles should remain relaxed during test.

OPPOSITION OF THUMB AND FIFTH FINGER

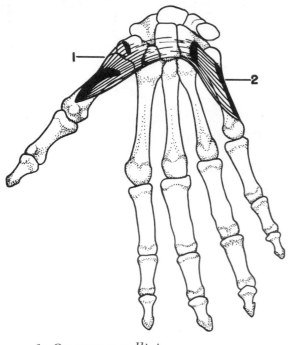

Range of Motion:

Pad of thumb should be brought flat against that of little finger with apposition of first and fifth metacarpal bones

Factors Limiting Motion:

1. Tension of transverse metacarpal ligament (limits motion of fifth metacarpal)
2. Tension of extensor tendons of first and fifth digits

Fixation:

Weight of forearm and hand

1. *Opponens pollicis*
2. *Opponens digiti minimi*

PRIME MOVERS

MUSCLE	ORIGIN	INSERTION
Opponens pollicis N: Median (C6, 7)	a. Ridge on trapezium bone b. Flexor retinaculum	a. Entire length of first metacarpal bone on radial side
Opponens digiti minimi N: Ulnar (C8, T1)	a. Convexity of hamulus of hamate bone b. Flexor retinaculum	a. Entire length of fifth metacarpal bone on ulnar side

Accessory Muscles

From the normal rest position the motion of thumb abduction must precede that of opposition. Therefore the Abductor pollicis longus and brevis muscles are in that sense accessory to the total motion.

OPPOSITION OF THUMB AND FIFTH FINGER

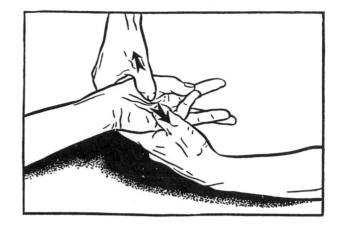

NORMAL AND GOOD

Sitting with hand resting palm upward on table.

Patient brings palmar surfaces of distal phalanges of thumb and fifth finger together.

The first and fifth metacarpals rotate toward the midline of the hand. The movement cannot be carried out by muscles other than the two opponens.

Resistance is given on distal end of first and fifth metacarpals on palmar surface with derotating pressure. The two muscles are graded separately.

FAIR AND POOR

Patient moves thumb and fifth finger through full range of opposition for fair grade and through partial range for poor grade.

The two muscles are graded separately. (Not illustrated.)

TRACE AND ZERO

The two muscles of opposition cannot be palpated unless the overlying superficial muscles are nonfunctioning.

★ ★ ★ ★ ★

Note: Flexor pollicis longus and Flexor pollicis brevis may draw thumb across hand toward fifth finger. This motion takes place in same plane as palm and should be distinguished from true opposition.

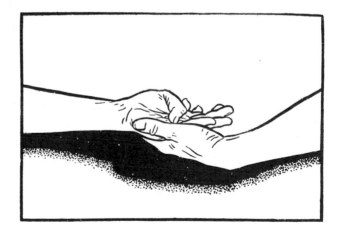

FOR NOTES:

MUSCLES OF THE FACE

In the testing of the face muscles, positioning is not a factor, and, with the exception of the muscles of mastication, only very fine movements are involved. Grades which may be used are: *zero*, if no contraction can be elicited; *trace*, for minimal muscle contraction; *fair*, for performance of the movement with difficulty; and *normal*, for completion of the movement with ease and control. Resistance may be given in the tests for the muscles of mastication.

MUSCLES OF THE FOREHEAD AND NOSE

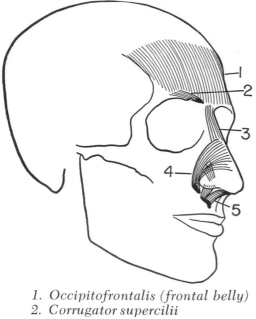

1. *Occipitofrontalis (frontal belly)*
2. *Corrugator supercilii*
3. *Procerus*
4. *Nasalis*
5. *Depressor septi*

PRIME MOVERS

MUSCLE	ORIGIN	INSERTION
Occipitofrontalis (frontal belly) N: Temporal branches of facial	Galea aponeurotica midway between coronal suture and orbital arch	Fibers are continuous medially with those of Procerus; intermediate fibers blend with Corrugator and Orbicularis oculi
Corrugator supercilii N: Temporal and zygomatic branches of facial	Medial end of superciliary arch	Deep surface of skin above middle of orbital arch
Procerus N: Buccal branches of facial	Fascia covering lower part of nasal bone and upper part of lateral nasal cartilage	Skin over lower forehead, between eyebrows
Nasalis N: Buccal branches of facial	Transverse part (compressor): Maxilla, above and lateral to incisive fossa Alar part (dilator): Greater alar cartilage	Thin aponeurosis continuous with muscle of opposite side Integument at point of nose
Depressor septi N: Buccal branches of facial	Incisive fossa of maxilla	Septum and back part of ala of nose

MUSCLES OF THE FOREHEAD AND NOSE

OCCIPITOFRONTALIS
(frontal belly)

Patient raises eyebrows, forming horizontal wrinkles in forehead (expression of surprise).

CORRUGATOR SUPERCILII

Patient draws eyebrows medially and downward, forming vertical wrinkles between brows (frowning).

PROCERUS

Patient lifts lateral borders of nostrils, forming diagonal wrinkles along bridge of nose (expression of distaste).

NASALIS

Patient dilates nostrils (alar part of Nasalis) followed by compression (transverse portion)

MUSCLES OF THE EYE

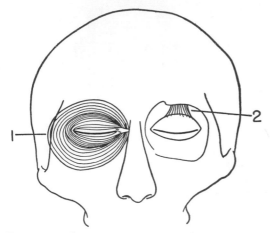

1. *Orbicularis oculi*
2. *Levator palpebrae superioris*

Diagram showing muscles used in conjugate ocular movements in the six cardinal directions of gaze

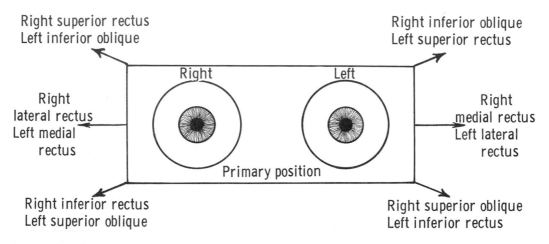

Right superior rectus
Left inferior oblique

Right inferior oblique
Left superior rectus

Right
lateral rectus
Left medial
rectus

Right
Left

Right
medial rectus
Left lateral
rectus

Primary position

Right inferior rectus
Left superior oblique

Right superior oblique
Left inferior rectus

From Chusid, J. G. Correlative Neuroanatomy and Functional Neurology, 14th ed. Los Altos, Calif., Lange Medical Publications, 1970, Page 94.

PRIME MOVERS

MUSCLE	ORIGIN	INSERTION
Orbicularis oculi N: Temporal and zygomatic branches of facial	Orbital part: a. Nasal part of frontal bone b. Frontal process of maxilla in front of lacrimal groove c. Anterior surface and borders of medial palpebral ligament	(Fibers form a complete ellipse without interruption, surrounding circumference of orbit and spreading over temple and downward on cheek)
	Palpebral part: Bifurcation of medial palpebral ligament	Lateral palpebral raphe
	Lacrimal part (tensor tarsi): posterior crest and adjacent part of lacrimal bone	Divides into two slips which insert into superior and inferior tarsi medial to puncta lacrimalia

(Continued on page 144.)

MUSCLES OF THE EYE

ORBICULARIS OCULI

Patient closes eyes tightly.

LEVATOR PALPEBRAE SUPERIORIS

Upper eyelids are lifted completely as eyes are turned upward.

RIGHT SUPERIOR RECTUS AND LEFT INFERIOR OBLIQUE

Patient moves eyes in a direction upward and to the right.

RIGHT SUPERIOR OBLIQUE AND LEFT INFERIOR RECTUS

Patient moves eyes in a direction downward and to the left.

The Rectus medialis and Rectus lateralis may be tested by movement of the eyes horizontally to the right and left. (Not illustrated.)

MUSCLE	ORIGIN	INSERTION
Levator palpebrae superioris N: Oculomotor	Inferior surface of small wing of sphenoid, superior and anterior to optic foramen	Forms broad aponeurosis which splits into 3 lamellae: superficial blends with upper part of orbital septum and is prolonged forward above Tarsalis superior to deep surface of skin of superior eyelid; middle into upper margin of Tarsalis superior; deepest into superior fornix of conjunctiva
Rectus superior N: Oculomotor	Superior part of fibrous ring surrounding optic foramen on superior, medial and inferior margins	Into sclera about 6 mm. behind cornea, on superior aspect of eyeball.
Rectus inferior N: Oculomotor	Inferior part of fibrous ring surrounding optic foramen on superior, medial and inferior margins	Into sclera about 6 mm. behind cornea
Rectus medialis N: Oculomotor	Medial part of fibrous ring surrounding optic foramen on upper, medial and lower margins	Into sclera on medial aspect of eyeball, about 6 mm. behind cornea.
Rectus lateralis N: Abducent	Two heads from lateral parts of bands surrounding optic foramen and adjoining part of orbital fissure	Into sclera on lateral aspect of eyeball, about 6 mm. behind cornea.
Superior oblique N: Trochlear	Above margin of optic foramen, from body of sphenoid	Passes forward, ending in tendon which plays in fibrocartilaginous pulley attached to trochlear fovea of frontal bone; tendon passes backward, lateralward, and downward to lateral aspect of eyeball, inserting into sclera behind the equator of the eyeball; thus muscle pulls in a forward, upward and medial direction
Inferior oblique N: Oculomotor	Orbital surface of maxilla, lateral to lacrimal groove	Passes lateralward, backward and upward to insert into lateral part of sclera somewhat posterior to insertion of Obliquus superior

144

FOR NOTES:

MUSCLES OF THE MOUTH

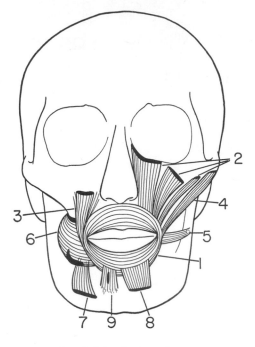

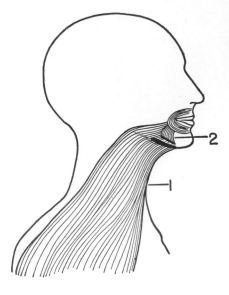

1. *Orbicularis oris*
2. *Zygomaticus minor*
3. *Levator anguli oris*
4. *Zygomaticus major*
5. *Risorius*
6. *Buccinator*
7. *Depressor anguli oris*
8. *Depressor labii inferioris*
9. *Mentalis*

1. *Platysma*
2. *Depressor anguli oris*

MUSCLE	ORIGIN	INSERTION
Orbicularis oris N: Buccal branches of facial	a. Fibers derived from other facial muscles, principally Buccinator, Levator anguli oris, and Depressor anguli oris b. Proper fibers of lips, from under surface of skin c. Fibers attached to maxilla and septum of nose above and to mandible below	a. Intermingling of transverse and oblique fibers comprising muscle b. Mucous membrane lining mouth cavity c. Decussation of some fibers of Buccinator at corner of mouth; Levator anguli oris fibers pass below, and Depressor anguli oris fibers pass above mouth
Zygomaticus minor N: Buccal branches of facial	Malar surface of zygomatic bone posterior to zygomaticomaxillary suture	Upper lip between angular head and Levator anguli oris Upper lip at corner of mouth

(Continued on page 149.)

MUSCLES OF THE MOUTH

ORBICULARIS ORIS
Patient approximates and compresses lips.

ZYGOMATICUS MINOR
Patient protrudes upper lip.

LEVATOR ANGULI ORIS
Patient lifts upper border of lip on one side without raising lateral angle of mouth (sneering). (Not Illustrated.)

ZYGOMATICUS MAJOR
Patient raises lateral angle of mouth upward and lateralward (smiling).

RISORIUS
Patient approximates lips and draws corners of mouth lateralward (grimacing).

MUSCLES OF MOUTH

BUCCINATOR

Patient approximates lips and compresses cheeks (blowing).

DEPRESSOR LABII INFERIOR

Patient protrudes lower lip (pouting).

DEPRESSOR ANGULI ORIS AND PLATYSMA

Patient draws corners of mouth downward strongly.

MENTALIS

Patient draws tip of chin upward. (Not illustrated.)

MUSCLE	ORIGIN	INSERTION
Levator anguli oris N: Buccal branches of facial	Canine fossa, immediately below infraorbital foramen	Angle of mouth, intermingling with Zygomaticus, Depressor anguli oris and Orbicularis
Zygomaticus major N: Buccal branches of facial	Zygomatic bone anterior to zygomaticotemporal suture	Angle of mouth, intermingling with Levator and Depressor anguli oris and Orbicularis oris
Risorius N: Mandibular and buccal branches of facial	Fascia over Masseter; muscle passes laterally superficial to Platysma	Skin at angle of mouth
Buccinator N: Buccal branches of facial	a. Outer surfaces of alveolar processes of maxilla above and mandible below, alongside the 3 molar teeth b. Pterygomandibular raphe	Fibers blend with deeper stratum of fibers in lips
Depressor anguli oris N: Mandible and buccal branches of facial	Oblique line of mandible	Angle of mouth
Depressor labii inferioris N: Buccal branches of facial	Oblique line of mandible, between symphysis and mental foramen	Skin of lower lip, blending with Orbicularis oris and opposite Depressor labii inferioris
Mentalis N: Mandibular and buccal branches of facial	Incisive fossa of mandible	Integument of chin
Platysma N: Cervical branch of facial	Fascia over superior Pectoralis major and Deltoideus muscles	a. Anterior fibers interlace with opposite muscle inferior and posterior to symphysis menti b. Posterior fibers insert into mandible below oblique line or blend with muscles near angle of mouth

MUSCLES OF MASTICATION

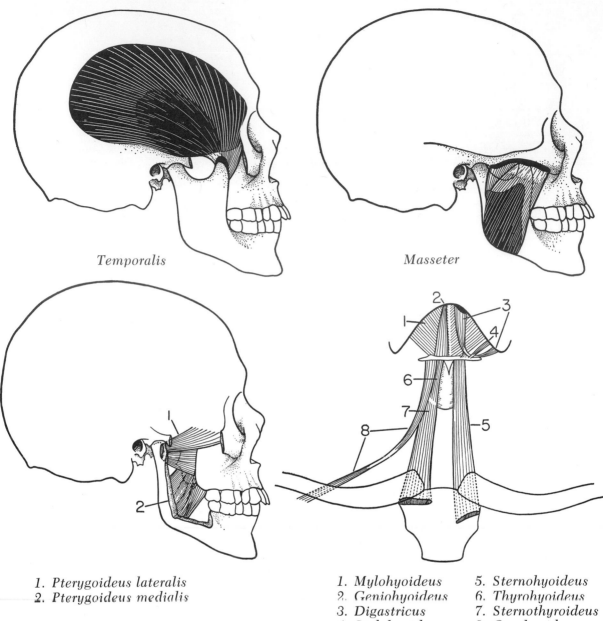

Temporalis

Masseter

1. *Pterygoideus lateralis*
2. *Pterygoideus medialis*

1. *Mylohyoideus*	5. *Sternohyoideus*
2. *Geniohyoideus*	6. *Thyrohyoideus*
3. *Digastricus*	7. *Sternothyroideus*
4. *Stylohyoideus*	8. *Omohyoideus*

PRIME MOVERS

MUSCLE	ORIGIN	INSERTION
Temporalis N: Deep temporal nerves from mandibular division of trigeminal	a. Temporal fossa b. Deep surface of temporal fascia	a. Medial surface, apex, and anterior border of coronoid process of mandible b. Anterior border of ramus of mandible nearly as far anterior as last molar tooth

(Continued on page 152.)

MUSCLES OF MASTICATION

TEMPORALIS, MASSETER AND PTERYGOIDEUS MEDIALIS

Patient closes jaws tightly.

PTERYGOID LATERALIS AND MEDIALIS (left)

Patient moves mandible laterally and forward to the right.

DIGASTRICUS AND SUPRAHYOID MUSCLES
(Hyoid bone is fixed by infrahyoid muscles.)
Patient depresses mandible.

Muscle	Origin	Insertion
Masseter N: Masseteric nerve from mandibular division of trigeminal	Superficial portion: a. Zygomatic process of maxilla b. Anterior two thirds of inferior border of zygomatic arch Deep portion: a. Posterior third of inferior border of zygomatic arch b. Whole medial surface of zygomatic arch	Angle and lower half of lateral surface of ramus of mandible a. Lateral surface of superior half of ramus of mandible b. Lateral surface of coronoid process
Pterygoideus lateralis N: Lateral pterygoid nerve from mandibular division of trigeminal	Superior head: a. Inferior part of lateral surface of great wing of sphenoid b. Infratemporal crest Inferior head: Lateral surface of lateral pterygoid plate	a. Depression in anterior part of neck of condyle of mandible b. Anterior margin of articular disk of temporomandibular articulation
Pterygoideus medialis N: Medial pterygoid nerve from mandibular division of trigeminal	a. Medial surface of lateral pterygoid plate b. Pyramidal process of palatine bone (second slip is lateral to Pterygoideus lateralis)	Inferior and posterior parts of medial surface of ramus and angle of mandibular foramen
Mylohyoideus N: Trigeminal	Whole length of mylohyoid line of mandible, from symphysis in front to last molar tooth	Body of hyoid bone
Geniohyoideus N: (C1) via hypoglossal	Inferior mental spine on inner surface of symphysis menti	Anterior surface of body of hyoid bone
Digastricus N: Posterior belly, facial; anterior belly, trigeminal	Posterior belly: mastoid notch of temporal bone Anterior belly: Depression on inner side of inferior border of mandible	(The two portions are united by an intermediate rounded tendon which perforates the Stylohyoideus muscle)
Stylohyoideus N: Facial	Styloid process near its base	Body of hyoid bone at junction with greater cornu, just above Omohyoideus muscle
Sternohyoideus N: (C1, 2, 3) via ansa cervicalis	a. Posterior surface of medial end of clavicle b. Posterior and superior part of manubrium	Inferior border of body of hyoid bone
Thyrohyoideus (C1, 2) via hypoglossal	Oblique line on lamina of thyroid cartilage	Inferior border of greater cornu of hyoid bone
Sternothyroideus N: (C1, 2, 3) via ansa cervicalis	Dorsal surface of manubrium below Sternohyoideus, from edge of cartilage of first and sometimes second rib	Oblique line on lamina of thyroid cartilage
Omohyoideus N: (C1, 2, 3) via ansa cervicalis	Inferior belly: cranial border of scapula, occasionally superior transverse ligament Superior belly: caudal border of body of hyoid bone	(The two portions are united by a central tendon held in position by a sheath of deep cervical fascia which is anchored to the clavicle and first rib)

II. SCREENING THE AMBULATORY PATIENT FOR MUSCLE TESTING BY GAIT ANALYSIS

BACKGROUND OF GAIT ANALYSIS

The subject of human gait has been of interest since man first experienced limitations in locomotion. In 1680 Borelli had the distinction of being the pioneer in a recorded approach to the problems of human movement including locomotion. Little progress was made until the first half of the nineteenth century, however, when the observational period in gait analysis began with the brothers Wilhelm and Edward Weber. Steindler summarizes their contribution as the observation and measurement of "alternation of swing and support, the inclination of the trunk in either phase, the relationship between the duration and the length of the step, and the rhythm of alternation in walking and running, setting up a special pattern for each. They also initiated investigations into the muscle effort involved in propulsion and restraint."

Visual recording, using photography and kymography, followed and permitted recording phases of gait. It was then possible to calculate velocity, acceleration and moving forces. The work of Braune and Fischer (1895) is the classical example of this approach.

More recently, study of individual muscles has permitted analysis of a pattern sequence of movement, first by palpation, later by electromyography. Morton and Schwartz added the concept and techniques of recording foot pressures.

The dynamic study of muscles encompassing the intensity and duration of muscle effort, exemplified by the University of California Prosthetics Devices Research studies, has provided still more tools for approaching the variations from normal gait so important to the clinician.

GAIT ANALYSIS

PHASES OF THE NORMAL PATTERN OF GAIT

There are two basic phases in the full cycle of a step: stance, the weightbearing period; and swing, the non-weightbearing portion of the step. In research on gait, a number of subdivisions of the two phases have been developed for study in detail, for example, heel-strike, foot-flat, early stance, mid-stance, heel-off, toe-off. However, for the purpose of a quick analysis of the patient's pattern of walking, four points in the cycle of a single step have been selected for easy identification and for recording any deviations from the normal gait. They are, in succession:

> Heel-strike
> Mid-stance
> Push-off
> Mid-swing

Elements of the Normal Pattern of Gait

Alignment
  1. Head is erect.
  2. Shoulders are level.
  3. Trunk is vertical.
Gross Movements
  1. Arms swing reciprocally and with equal amplitude at normal walking speed.
  2. Steps are of the same length and timing is synchronized.
  3. Body undergoes vertical oscillations that are definite and have an even tempo.
Fine Movements
  1. Pelvis undergoes slight
     a. transverse rotation: rotation is medial from the end of push-off to mid-

stance and lateral from mid-stance through push-off.

b. anterior-posterior rotation (tipping): forward inclination of the pelvis is maintained during the complete cycle. The maximum amount of anterior rotation is evident before heel-strike and the minimum before mid-stance (excursion 3–5 degrees);

c. lateral tilt: maximum tilt is downward at mid-swing on the side of the swinging leg; and

d. lateral displacement: maximum displacement is lateral at mid-stance on the side of the weightbearing leg.

2. Legs rotate slightly medially at hip and knee during swing and heel-strike to near mid-stance, followed by a change to lateral rotation which continues through push-off.

3. Knees have two alternations of extension and flexion in a single cycle:

a. extension of knee at heel-strike (not locked or tightly extended);

b. slight flexion following heel-strike (continuing through mid-stance);

c. extension following mid-stance; and

d. flexion during push-off and swing.

4. Ankles

a. rotate forward in an arc about a radius formed by the heel at heel-strike and around a center in the forefoot at push-off; and

b. display maximum dorsiflexion at the end of stance phase and maximum plantar-flexion at the end of push-off.

CAUSES OF DEVIATIONS IN GAIT

There are many factors which cause deviations in the normal pattern of walking. The most common are:

1. Pain or discomfort during weightbearing or movement;

2. Muscle weakness;

3. Limitation of joint motion (often with muscle shortening);

4. Incoordination of movement; and

5. Changes in bone or soft tissue (including amputations).

Pain or discomfort can bring about distortions in the gait pattern varying from minor changes in alignment or movement to extreme deviations. Before observations are recorded, these factors should be explored by questioning the patient and noting protective responses. Clues such as shortening the stance (weightbearing phase) or facial expression are of particular importance if the patient is unable to respond to questions.

Muscle weakness may be moderate and generalized, resulting in a broadened base of support, short steps, diminished arm swing, and difficulty in balance. In other instances, there may be extensive weakness in certain muscle groups but sufficient strength in others to allow ambulation. The extreme deviations are usually seen in the latter group.

Limitation of joint motion is most commonly found as a result of a pathological condition, such as arthritis; surgical procedures, for example the insertion of metal implants; and disuse of a part or of the whole body from a variety of causes. Deviations in gait due to joint limitation can be identified in the analysis and verified by joint measurements.

Incoordination resulting from neuropathological conditions (e.g., spastic cerebral palsy, hemiparesis secondary to cerebral vascular accident, Parkinson's syndrome), often leads to distinctive patterns of walking that can readily be identified and a description recorded. Hypertonic states are characterized by the lack of ability to activate muscle groups selectively and to combine them into the various patterns required for normal walking. The patient tends to respond with a total flexion or a total extension pattern when moving the limbs.

Deformities of bone and soft tissue lead to a variety of deviations in gait. Examples of such deformities are bone shortening following a fracture, congenital malformations, or dense scar tissue from a severe burn.

OBJECTIVES OF GAIT ANALYSIS

The objectives of gait analysis are to identify deviations and obtain information that may assist in determining the cause of the deviations and provide a basis for the use of therapeutic procedures or supportive devices to improve the walking pattern. Muscle tests can be used to determine the level of muscular weakness. However, by providing a gait analysis to screen ambulatory patients, it is possible to shorten and simplify the test procedures. Apparent areas of weakness found in the gait analysis may be validated by the results of the muscle tests and the accuracy of the tests supported by the findings in the analysis.

PROCEDURE

A special sheet may be devised for recording the results of the gait analysis for patients with extensive involvement. Headings can be used such as Gait Deviation, Cause of Deviation, and Corrective Procedures. A patient's disability, however, is frequently limited to one area and the information can be recorded without the use of a separate sheet.

The patient, if possible, should walk at a speed that is considered by the examiner to be normal for his age. A very slow gait tends to mask deviations as the steps are shortened and ranges of motion minimized. Such a gait may indicate generalized weakness with instability. If incoordination appears to be a problem, the ability to produce joint motion at hip, knee and ankle in various combinations may be tested in order to determine the particular combination, or combinations, lacking for a normal gait.

After deviations in gait resulting from pain, generalized weakness, incoordination and fixed deformities have been screened out, the alignment of the body and gross movements should be reviewed. Any variation from the normal pattern indicates the need for a consideration of the fine movements with attention to each segment in sequence. If deviations are present, they are noted and follow-up muscle tests or other evaluative procedures instituted as indicated.

The next section is devoted to an analysis of (1) the four phases of normal gait with lateral and anterior or posterior views to illustrate each; and (2) the common deviations in gait with muscles that may be tested for each deviation. The range of motion to be checked is included where limitation is commonly the cause of the deviation.

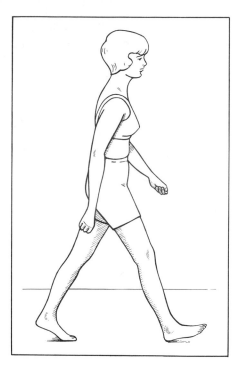

GAIT PHASE I: HEEL-STRIKE

NORMAL PATTERN FROM *LATERAL* VIEW
(Right Leg)

1. Head and trunk are vertical. (Right arm is posterior to midline of body with elbow extended; left arm is anterior with elbow partially flexed.)
2. Pelvis has slight anterior rotation.
3. Right knee is extended.
4. Right foot is approximately at a right angle to leg.

Common Deviations from Normal

*1. Head and trunk are shifted forward at heel-strike.
2. Pelvis has posterior rotation.

3. Knee is in locked extension or hyper-extension.
4. Foot is placed flat on floor in plantar flexion. There may be a slapping of forefoot.

Test the Following Muscles

1. Knee extensors.

2. Back extensors and hip flexors (check range of motion in hip flexion).

3. Knee extensors and flexors.

4. Ankle dorsiflexors.

---

*Places center of gravity anterior to knee joint to prevent knee flexion.

GAIT PHASE I: HEEL-STRIKE

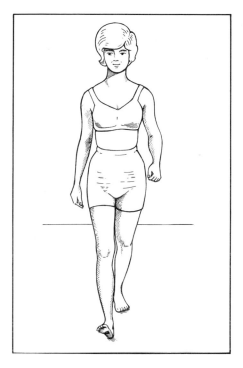

NORMAL PATTERN FROM *ANTERIOR* VIEW
(Right Leg)

1. Head and trunk are vertical. (Arms swing at an equal distance from body.)
2. Leg is in vertical alignment with pelvis.
3. Plantar surface of forefoot is visible.

Common Deviations from Normal

*1. Trunk is displaced to right and leg is in lateral rotation at hip (step is shortened).
2. Leg is in abduction at hip.
3. Plantar surface of forefoot is not visible.

Test the Following Muscles

1. Hip medial rotators, knee extensors and evertors of foot.
2. Hip adductors.
3. Ankle dorsiflexors.

---

*If knee extensors are weak, leg may be placed in lateral rotation at the hip to prevent flexion. If foot evertors are weak, lateral rotation prevents ankle from rolling outward.

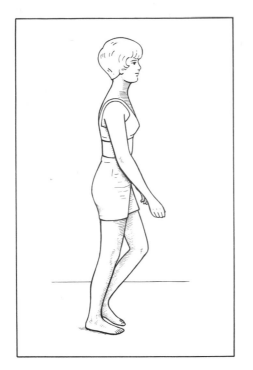

GAIT PHASE II: MID-STANCE

NORMAL PATTERN FROM *LATERAL* VIEW
(Right Leg)

1. Head and trunk are vertical. (Arms are near midline of body, elbows partially flexed.)
2. Pelvis has very slight anterior rotation.
3. Right knee is in slight flexion.

Common Deviations from Normal

*1a. Head and trunk are shifted forward at the hip joint with exaggerated anterior rotation of pelvis.

**b. Head and trunk are shifted backward at the hip joint with posterior rotation of pelvis.

2. Pelvis has exaggerated anterior rotation.

3a. Knee is in extension or hyperextension.

b. Knee has exaggerated flexion.

Test the Following Muscles

1a. Knee extensors.

b. Hip extensors.

2. Abdominals and hip extensors (check range of motion in hip extension).

3a. Knee flexors and extensors and ankle dorsiflexors (check range of motion in dorsiflexion of ankle).

b. Ankle plantar flexors.

---

*Places center of gravity anterior to knee joint to prevent knee flexion.

**Places center of gravity posterior to hip joint to prevent forward shift of trunk.

GAIT PHASE II: MID-STANCE

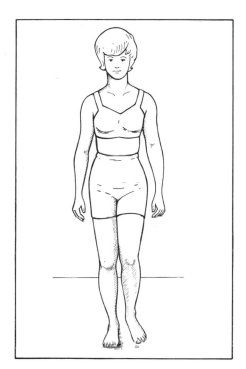

NORMAL PATTERN FROM *ANTERIOR* VIEW
(Right Leg)

1. Head and trunk are vertical. (Arms are at an equal distance from body.)

2. Pelvis is tilted downward very slightly on left side.

3. Leg is in slight lateral rotation at hip.

Common Deviations from Normal

*1. Head and trunk tip to right and pelvis tilts *upward* on left side. Right arm is away from body (Gluteus Medius Gait).

2. Pelvis has an exaggerated *downward* tilt on left side (Trendelenberg gait).

**3. Leg is in exaggerated outward rotation at hip.

Test the Following Muscles

1. *Right* hip abductors.

2. *Right* hip abductors.

3. Hip adductors and medial rotators, extensors of knee, and evertors of ankle.

---

*If bilateral, patient tips to each side alternately with "waddle" gait.
**See note on extensors of knee and evertors of ankle in heel strike, anterior view, on page 157.

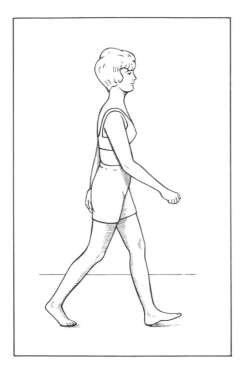

GAIT PHASE III: PUSH-OFF

NORMAL PATTERN FROM *LATERAL* VIEW
(Right Leg)

1. Right arm is anterior to midline of body with elbow partially flexed, left arm is posterior with elbow extended.
2. Pelvis is in anterior rotation.
3. Knee is slightly flexed.
4. Ankle is plantar flexed.
5. Toes are in hyperextension at metatarsophalangeal joint.

Common Deviations from Normal

*1. Arms are at an uneven distance from midline of body with both elbows flexed.
2. Pelvis has exaggerated anterior rotation.
3. Knee is partially flexed.
4. Plantar flexion is limited and ankle may be in dorsiflexion.
5. Metatarsophalangeal joints are straight.

Test the Following Muscles

1. Ankle plantar flexors and hip and knee extensors.
2. Abdominals and hip extensors (check range of motion in hip extension).
3. Ankle plantar flexors.
4. Same as Number 3.

5. Same as Number 3 (check range of motion in hyperextension).

---

*An exaggerated forceful arm swing is used to assist in push-off and step may be shortened.

160

GAIT PHASE III: PUSH-OFF

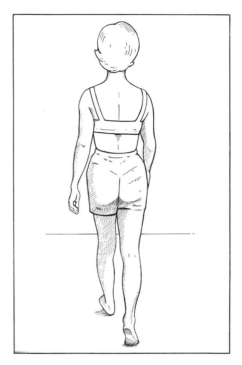

NORMAL PATTERN FROM *POSTERIOR* VIEW
(Right Leg)

1. Arms are at an equal distance from body, right elbow partially flexed, and left extended.

2. Leg is slightly laterally rotated at hip.

3. Plantar surface of heel and mid-foot are visible and forefoot is in contact with floor.

Common Deviations from Normal

*1. Arms are at an unequal distance from body with both elbows flexed.

**2. Leg is in exaggerated lateral rotation at hip.

3. Plantar surface of foot is not visible. Forefoot is not in contact with floor as heel is lifted.

Test the Following Muscles

1. Ankle plantar flexors and hip and knee extensors.

2. Same as Number 1.

3. Same as Number 1.

---

*See footnote on page 160.
**Knee may be forcefully extended to assist in push-off.

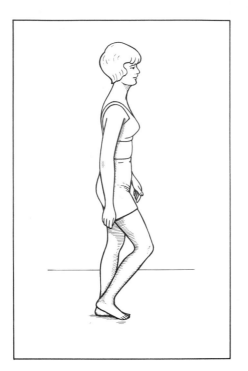

GAIT PHASE IV: MID-SWING

NORMAL PATTERN FROM *LATERAL* VIEW
(Right Leg)

1. Arms are near midline of body.
2. Pelvis has very slight anterior rotation.
3. Hip and knee are flexed.
4. Foot is at right angle to leg.

Common Deviations from Normal

1. Pelvis has posterior rotation.

2. Hip and knee flexion are exaggerated and forefoot is dropped (Steppage Gait).
3. Toe drags on floor.

Test the Following Muscles

1. Back extensors and hip flexors (check range of motion in hip flexion).
2. Ankle dorsiflexors.

3. Hip flexors, knee flexors and ankle dorsiflexors.

GAIT PHASE IV: MID-SWING

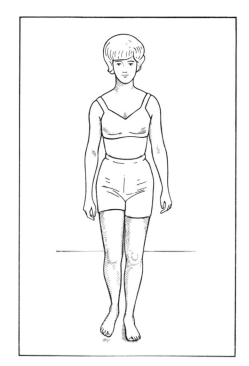

NORMAL PATTERN FROM *ANTERIOR* VIEW
(Right Leg)

1. Head and trunk are vertical.
2. Arms are at an equal distance from body.
3. Pelvis is tilted slightly downward on right side.
4. Leg is in vertical alignment with pelvis and has slight medial rotation at hip.
5. Foot is at right angle to leg with slight eversion.

Common Deviations from Normal

1. Trunk is displaced to left. Pelvis is lifted on right side (hip hiking).
*2. Leg is in abduction.

3. Leg is laterally rotated at hip.
4. Forefoot is dropped, eversion is not visible.

Test the Following Muscles

1. Hip and knee flexors and ankle dorsiflexors.
2. Same as Number 1 (check range of motion in hip adduction and flexion and knee flexion).
3. Hip medial rotators and foot evertors.
4. Ankle dorsiflexors and foot evertors.

---

*Leg may be circumducted (an arc of abduction) through swing phase.

REFERENCES

Bailey, J. C. Manual Muscle Testing in Industry. Phys. Ther. Rev., *41*:165–169, 1961.

Basmajian, J. V. Muscles Alive: Their Functions Revealed by Electromyography. 2nd ed. Baltimore, Williams and Wilkins Company, 1967.

Basmajian, J. V. Recent Advances in the Functional Anatomy of the Upper Limb. Amer. J. Phys. Med., *48*:165–177, 1969.

Beasley, W. C. Influence of Method on Estimates of Normal Knee Extensor Force Among Normal and Postpolio Children. Phys. Ther. Rev., *36*:21–41, 1956.

Bennett, R. L. Muscle Testing: A Discussion of the Importance of Accurate Muscle Testing. Physiotherapy Rev., *27*:242–243, 1947.

Borden, R., and Colachis, S. C. Quantitative Measurement of the Good and Normal Ranges in Muscle Testing. Phys. Ther., *48*:839–843, 1968.

Borelli, J. A. *De Motu Animalium.* Rome, 1680.

Braune, C. W., and Fisher, O. Der Gang des Menschen. I Teil. Versuche unbelasten und belasten Menschen. Abhandl. d. Math.-Phys. Cl. d. k. Sachs. Gesellsch. Wissensch., *21*:153–322, 1895.

Brunnstrom, S. Muscle Group Testing. Physiotherapy Rev., *21*:3–21, 1941.

Brunnstrom, S., and Dennen, M. Round Table on Muscle Testing. Annual Conference of American Physiotherapy Association, Federation of Crippled and Disabled, Inc., New York, 1931, pp. 1–2 (mimeographed).

Carroll, R. L. Rate and Amount of Improvement in Muscle Strength Following Infantile Paralysis. Physiotherapy Rev., *22*:243–257, 1942.

Chusid, J. G. Correlative Neuroanatomy and Functional Neurology. 14th ed. Los Altos, Calif., Lange Medical Publications, 1970, p. 94.

Close, J. R. Motor Function in the Lower Extremity. Springfield, Illinois, Charles C Thomas, 1964.

Close, J. R., and Todd, F. N. The Phasic Activity of the Muscles of the Lower Extremity and the Effect of Tendon Transfer. J. Bone Joint Surg., *41-A*:189–208, 1959.

Cunningham, D. C. Components of Floor Reactions During Walking. Prosthetic Devices Research Project, Institute of Engineering Research, University of California, Berkeley. Series II, Issue 14, November 1950 (Reissued October 1958).

Downer, A. H. Strength of the Elbow Flexor Muscles. Phys. Ther. Rev., *33*:68–70, 1953.

Ducroquet, R., Ducroquet, J., and Ducroquet, P. Walking and Limping: A Study of Normal and Pathological Walking. Philadelphia, J. B. Lippincott Company, 1965. English Translation, 1968.

Eberhart, H. D., Inman, V. T., and Bresler, B. The Principal Elements in Human Locomotion. *In:* Klopsteg, P. E., and Wilson, P. D. Human Limbs and Their Substitutes, Chapter 15. New York, McGraw-Hill Book Company, 1954.

Eberhart, H. D., Inman, V. T., Saunders, J. B. DeC. M., Levens, A. S., Bresler, B., and McCowan, T. D. Fundamental Studies of Human Locomotion and Other Information Relating to the Design of Artificial Limbs. A Report to the National Research Council, Committee on Artificial Limbs, University of California, Berkeley, 1947.

Elftman, H. A Cinematic Study of the Distribution of Pressure in the Human Foot. Anat. Rec., 59: 481–491, 1934.

Elftman, H. The Basic Pattern of Human Motion. Ann. N.Y. Acad. Sci., 5:1207–1212, 1950.

Fisher, F. J., and Houtz, S. J. Evaluation of the Function of the Gluteus Maximus Muscle. Amer. J. Phys. Med., 47:182–191, 1968.

Gonnella, C. The Manual Muscle Test in the Patient's Evaluation and Program for Treatment. Phys. Ther. Rev., 34:16–18, 1954.

Gonnella, C., Harmon, G., and Jacobs, M. The Role of the Physical Therapist in the Gamma Globulin Poliomyelitis Prevention Study. Phys. Ther. Rev., 33:337–345, 1953.

Granger, C. V. The Clinical Discernment of Muscle Weakness. Arch. Phys. Med., 44:430–438, 1963.

Gray, H. Anatomy of the Human Body. 28th ed. Edited by C. M. Goss. Philadelphia, Lea and Febiger, 1966.

Green, D. R. The Effects of Aging on the Component Movements of Human Gait. Doctoral Dissertation, University of Wisconsin, 1959.

Green, D. L., and Morris, J. M. Role of the Adductor Longus and Magnus in Postural Movements and in Ambulation. Amer. J. Phys. Med., 19:223–239, 1970.

Grossiord, A., and Husson, A. Le pronostic musculaire dans la poliomyélite. Éléments cliniques d'appréciation. Sem. Hôp. Paris, 291:271–278, 1953.

Hammon, W. McD., Coriell, L. L., and Stokes, J., Jr. Evaluation of Red Cross Gamma Globulin as a Prophylactic Agent for Poliomyelitis. I. Plan of Controlled Field Tests and Results of 1951 Pilot Study in Utah. J.A.M.A., 150:739–749, 1952.

Hammon, W. McD., Coriell, L. L., and Stokes, J., Jr. Evaluation of Red Cross Gamma Globulin as a Prophylactic Agent for Poliomyelitis. 2. Conduct and Early Follow-up of 1952 Texas and Iowa-Nebraska Studies. J.A.M.A., 150:750–756, 1952.

Hines, T. F. Manual Muscle Examination. Chapter 8, Therapeutic Exercise, Edited by S. Licht. New Haven, E. Licht, Publisher, 1961.

Hubbard, A. W., and Stetson, R. H. An Experimental Analysis of Human Locomotion. Amer. J. Physiol., 124:300–314, 1938.

Iddings, D. M., Smith, L. K., and Spencer, W. A. Muscle Testing: Part 2. Reliability in Clinical Use. Phys. Ther. Rev., 41:249–256, 1961.

Inman, V. T. Functional Aspects of the Abductor Muscles of the Hip. J. Bone Joint Surg., 29:607–619, 1947.

Jarvis, D. K. Relative Strength of Hip Rotator Muscle Groups. Phys. Ther. Rev., 32:500–503, 1952.

Kendall, F. P. Testing the Muscles of the Abdomen. Physiotherapy Rev., 21:22–24, 1941.

Kendall, H. O., and Kendall, F. P. Care During the Recovery Period in Paralytic Poliomyelitis. U.S. Public Health Bull. No. 242, revised 1939.

Kendall, H. O., and Kendall, F. P. Muscles, Testing and Function. 1st ed. Baltimore, Williams & Wilkins Company, 1949.

LaBau, M. M., Raptou, A. D., and Johnson, E. W. Electromyographic Study of Function of Iliopsoas Muscle. Arch. Phys. Med., 46:676–679, 1965.

Legg, A. T. The Early Treatment of Poliomyelitis and the Importance of Physical Therapy. J.A.M.A., 107:633–645, 1936.

Liberson, W. T. Biomechanics of Gait: A Method of Study. Arch. Phys. Med., 46:37–48, 1965.

Lilienfeld, A. M., Jacobs, M., and Willis, M. A Study of the Reproducibility of Muscle Testing and Certain Other Aspects of Muscle Scoring. Phys. Ther. Rev., 34:279–289, 1954.

Lovett, R. W. The Treatment of Infantile Paralysis. 2nd ed. Philadelphia, P. Blakiston's Son & Co., 1917, p. 136.

Lovett, R. W., and Martin, E. G. Certain Aspects of Infantile Paralysis and a Description of a Method of Muscle Testing. J.A.M.A. 66:729–733, 1916.

Lowman, C. L. A Method of Recording Muscle Tests. Amer. J. Surg., New Series, 3:588–591, 1927.

Lowman, C. L. Relation of Abdominal Muscles to Paralytic Scoliosis. J. Bone Joint Surg., 14:763–772, 1932.

Lowman, C. L. Muscle Strength Testing. Physiotherapy Rev., 20:69–71, 1940.

Lowman, C. L., and Roen, S. G. Therapeutic Use of Pools and Tanks. Philadelphia, W. B. Saunders Company, 1952.

MacConaill, M. A., and Basmajian, J. V. Muscles and Movements, a Basis for Human Kinesiology. Baltimore, Williams & Wilkins Company, 1969.

McFarland, J. W., and Graves, D. Grading of Spasm in Infantile Paralysis. Arch. Phys. Med., 25:553–556, 1944.

Medical Research Council. Aids to the Investigation of Peripheral Nerve Injuries. Nerve Injuries Committee, War Memorandum No. 7. London, H.M. Stationery Office, 1942.

Morton, D. J., and Fuller, D. D. Human Locomotion and Body Form. Baltimore, Williams and Wilkins Company, 1952.

Murray, M. P. Gait as a Total Pattern of Movement. Amer. J. Phys. Med., 46:290–333, 1967.

Murray, M. P., and Clarkson, B. H. The Vertical Pathways of the Foot During Walking: I. Range of Variability in Normal Men. J. Amer. Phys. Ther. Ass., 46:585–589, 1966.

Murray, M. P., Drought, A. B., and Kory, R. C. Walking Patterns of Normal Men. J. Bone Joint Surg., 46–A:335–360, 1964.

Murray, M. P., Kory, R. C., Clarkson, B. H., and Sepic, S. B. A Comparison of Free and Fast Speed Walking Patterns of Normal Men. Amer. J. Phys. Med., 43:8–24, 1966.

Ogg, H. L. Measuring and Evaluating the Gait Patterns of Children. J. Amer. Phys. Ther. Ass., 43:717–720, 1963.

Partridge, M. J., and Walters, C. E. Participation of the Abdominal Muscles in Various Movements of the Trunk in Man: An Electromyographic Study. Phys. Ther. Rev., 39:791–800, 1959.

Patek, S. D. The Angle of Gait in Women. Amer. J. Phys. Anthrop., 9:273–291, 1926.

Perry, J. A Clinical Interpretation of the Mechanics of Walking. Phys. Ther. 47:778–801, 1967.

Plastridge, A. L. Gaits. Physiotherapy Rev., 21:24–29, 1941.

Plastridge, A. L. Round Table on Infantile Paralysis. Annual Conference of American Physiotherapy Association. Palo Alto, Stanford University Press, 1942. p. 47.

Pocock, G. S. Electromyographic Study of the Quadriceps During Resistive Exercise. J. Amer. Phys. Ther. Ass., 43:427–434, 1963.

Reeder, T. Electromyographic Study of the Latissimus Dorsi Muscle. J. Amer. Phys. Ther. Ass., 43:165–172, 1963.

Rodenberger, M. L. Method of Recording Progress in Gait Training. Phys. Ther. Rev., 30:92–94, 1950.

Saunders, J. B. DeC. M., Inman, V. T., and Eberhart, H. W. The Major Determinants in Normal and Pathological Gait. J. Bone Joint Surg., 35–A: 543–558, 1953.

Schochrin, W. A. Über die Beständigkeit des Altersverhältnisses Zwischen der Muskelkraft der Strecker und Beuger. Arbeitsphysiol., 8:607–609, 1935.

Schwartz, R. P., Heath, A. L., Misiek, W., and Wright, J. N. Kinetics of Human Gait. J. Bone Joint Surg., 16:343–350, 1934.

Schwartz, R. P., Heath, A. L., Morgan, D. W., and Towne, R. C. A Quantitative Analysis of Recorded Variables in the Walking Patterns of "Normal" Adults. J. Bone Joint, Surg. 46:324–334, 1964.

Scott, M. G. Analysis of Human Motion. A Textbook in Kinesiology. New York, F. S. Crofts & Co., 1942.

Sheffield, F. J. Electromyographic Study of the Abdominal Muscles in Walking and Other Movements. Amer. J. Phys. Med., 41:142–147, 1962.

Smith, L. K., Iddings, D. M., Spencer, W. A., and Harrington, P. R. Muscle Testing: Part 1. Description of a Numerical Index for Clinical Research. Phys. Ther. Rev., 41:99–105, 1961.

Steindler, A. Historical Review of the Studies and Investigations made in Relation to Human Gait. J. Bone Joint Surg., 35–A:540–543, 1953.

Sutherland, D. H. An Electromyographic Study of the Plantar Flexors of the Ankle in Normal Walking on the Level. J. Bone Joint Surg., 48–A:66–71, 1966.

Ufland, J. M. Einfluss des Lebensalters, Geschlechts, der Konstitution und des Berufs auf die Kraft Verschiedener Muskelgruppen; über den Einfluss des Lebensalters auf die Muskelkraft. Arbeitsphysiol., 6:653–663, 1933.

Walters, C. C., and Hoth, M. Aid for Recording Gross Evaluation. Phys. Ther., 44:179, 1964.

Weber, W., and Weber, E. Mechanik der menschlichen Gehwerkzeuge. Gottingen, Dietrich, 1836.

Williams, E. T., Top, F. T., and Suchomel, L. Graphic Method for Rapid Estimation of Clinical Status in Poliomyelitis. Arch. Phys. Med., 27:430–436, 1946.

Williams, M. Manual Muscle Testing: Development and Current Use. Phys. Ther. Rev., 36:797–805, 1956.

Williams, M., and Lissner, H. R. Biomechanical Analysis of Knee Function. J. Amer. Phys. Ther. Ass., 43:93–99, 1963.

Williams, M., and Lissner, H. R. Biomechanics of Human Motion. Philadelphia, W. B. Saunders Company, 1962.

Williams, M., and Stutzman, L. Strength Variation Through the Range of Joint Motion. Phys. Ther. Rev., 39:145–152, 1959.

Wintz, M. M. Variations in Current Muscle Testing. Phys. Ther. Rev., 39:466–475, 1959.

Wright, W. G. Muscle Training in the Treatment of Infantile Paralysis. Boston, M. & S. J., 167:567–574, 1912.

Zausmer, E. Evaluation of Strength and Motor Development in Infants. Phys. Ther. Rev., 33:575–581, 621–629, 1953.